LINEAR IC

principles, experiments, and projects

by

Edward M. Noll

Howard W. Sams & Co., Inc.
4300 WEST 62ND ST. INDIANAPOLIS, INDIANA 46268 USA

FIRST EDITION

SECOND PRINTING—1977

International Standard Book Number: 0-672-21019-3
Library of Congress Catalog Card Number: 73-90294

Preface

The integrated circuit is a natural follow-up to the discrete semiconductor device. In the integrated circuit, various semiconductor devices such as diodes, bipolar transistors, field-effect transistors, etc., are incorporated into a single semiconductor package. Discrete transistors are indeed small devices in themselves, but the integrated circuit permits an even greater number of semiconductor devices to be packaged together in such a manner that they can internally perform complete simple and complex electronic functions.

ICs are now produced in great numbers and can be found in most electronic equipment. They are a part of most home-entertainment units coming off the production lines today. These units include radio and television receivers, audio amplifiers, a-m and fm systems, two-way rados, etc. The little electronic calculator you can hold in your hand is just one example of the almost unbelievable extent to which integrated circuits can miniaturize electronic equipment. Commercial and industrial electronic equipment also employ these little magic devices.

When these devices first became available, many thought that the need for capable electronic technicians and engineers would decline because electronic equipment would now be automatic. The same fate was predicted for electronic experimentation. How wrong those beliefs were.

What the integrated circuit did do was to permit more compact miniaturization and make feasible more complex functions. It triggered the need for a greater depth of electronic knowledge and an even greater experimentation effort to find out how many ways the IC could perform a service. Knowledge was not turned off, nor was experimentation stymied. Instead, the integrated circuit came to be realized as an exciting device and an essential part in many electronic systems.

EDWARD M. NOLL

Contents

1

Basic Semiconductor Principles

The technician or engineer who works with integrated circuits has a two-fold responsibility. He must know how the device functions in the system with which it is associated, as well as have some understanding of what takes place within the device. (There are arguments occasionally with regard to the latter.) However, as the device becomes increasingly complex, it becomes more difficult to know how to apply the device or to understand its multiple-function capabilities in a system. Therefore, it is best that you know integrated circuits, or ICs as they are usually called, inside and outside.

Integrated circuits are now being produced in the millions. In varying degrees, these tiny packages of electronic marvels have infiltrated all facets of the electronics industry. Inside each package are components numbering from a few to hundreds of items. One device may function in a routine way in a simple system; another will perform a complex function in a complex electronic system. The device may be just a simple amplifier or a complete system.

Modern electronic entertainment units use their share of integrated circuits, but commercial and industrial electronic systems have come to depend on these tiny devices. Test equipment (Fig. 1-1) and two-way radio gear of all types (Fig. 1-2) have led the way in using ICs to advantage.

BASIC IC STRUCTURES

The two basic integrated circuit structures are monolithic and hybrid. In the monolithic type, the entire integrated circuit is completed as a single silicon die. All circuit components are an inherent part of the structure, formed within or on top of a tiny block of silicon. All

Courtesy Heath Co.

Fig. 1-1. A test instrument utilizing integrated circuits.

components are inseparable in a continuous array of silicon atoms—of proper polarization and differing degrees of impurity. Interconnections between components and connections to the output leads are handled by metalized patterns included in the manufacturing process. The photomicrograph of Fig. 1-3 shows an integrated circuit and its terminals. Thin wire leads connect the terminals to the pins of the integrated circuit casing.

The structure of a monolithic IC is built around a single silicon specimen. More than one monolithic device can be incorporated in a single case. Electrical isolation among the various devices and components is handled by polarization of the various layers and the deposition of special insulating strata. This latter process is called multiphasing (Fig. 1-4). Isolating dielectric barriers separate individual monolithic components or groups of components both electrically and physically.

In a hybrid-type IC, metallic depositions or wire bonds interconnect very tiny active and discrete components within a small case,

Courtesy Narco Avionics.

Fig. 1-2. Integrated circuits are used in this compact aviation communicator.

8

Fig. 1-3. Integrated circuit mounted in case.

Fig. 1-5. A hybrid circuit can contain one or several individual mono-lithic structures plus discrete components in its design.

These various IC constructions are discussed further in Chapter 2. Also considered are the various active and passive electronic components that can be incorporated into an integrated circuit. But before examining the construction of ICs, a review of certain solid-state

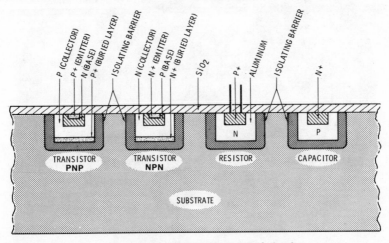

Fig. 1-4. Multiphased integrated circuits.

principles will assist in understanding IC structures and modes of operation.

SEMICONDUCTOR FUNDAMENTALS

The factor that differentiates one element from another is determined by the grouping, number, and placement of its electrons around the atomic nucleus. Electrons may occupy one or more energy bands or shells. These shells are spaced a certain distance from the nucleus

(A) IC showing external toroid coils.

(B) Multichip LC.

Fig. 1-5. Hybrid integrated circuits.

and contain a specific number of electrons of approximately the same energy level. Those electrons in orbit near the nucleus have a low energy level and are tightly held to the atomic structure. Electrons in the outer shell are of much higher energy content and are held less tightly.

The Silicon Atom

The element silicon is basic to most diodes, transistors, and integrated circuits in practical use today. The makeup of the silicon atom is illustrated in Fig. 1-6A. It consists of 2 electrons in the first shell

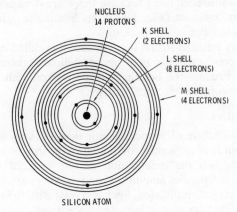

(A) Silicon atom showing energy bands.

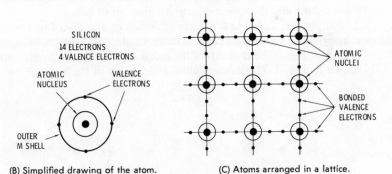

(B) Simplified drawing of the atom.   (C) Atoms arranged in a lattice.

Fig. 1-6. Atomic structure of silicon.

(K), 8 electrons in the second shell (L), and 4 electrons in the outer shell (M). Customarily, the simplified version shown in Fig. 1-6B is employed and shows the nucleus and the 4 electrons of the outer shell. It is the outer, so-called valence electrons, that determine the electrical characteristics of the particular element.

11

There are never more than 8 electrons in the outer shell of an element, but there may be fewer. When there are exactly 8 electrons, an element has a high stability because the valence electrons are bound tightly to the atom. Such an element serves as an excellent electrical insulator. Fewer than 8 valence electrons results in a less stable atom. Atoms having 5, 6, and 7 electrons tend to borrow additional electrons from other atoms. Those with 1, 2, or 3 valence electrons lose electrons to other atoms.

The interlocking of valence electrons among atoms produces stable molecules of a substance, forming crystalline formations of molecules. Silicon is an unusual element because there are 4 valence electrons in a balanced arrangement (Fig. 1-6C). Its binding is tight, and it takes on the characteristics of an insulator.

When highly purified silicon is produced, it is called *intrinsic* silicon. Although its conductivity is poor, its crystalline makeup is not a perfect insulator because the bonding can be broken with high temperature or an appropriate electrical energy.

Adding Impurities

In the manufacture of semiconductor devices, very small and very accurate amounts of impurities are added to the intrinsic crystal. These impurities establish the polarization of the material and lower its resistivity by a specific amount, giving a very carefully regulated amount of electrical conduction. Such a crystal with the proper amount of *doping* (added impurity) is called an *extrinsic* semiconductor. It has a conductivity that lies somewhere between the high conductivity of a conductor and the low conductivity of an insulator.

The impurities added to the intrinsic silicon crystal are of two types: one has only 3 electrons in the valence shell, while the other has 5 valence electrons. The makeup of the more common impurity elements is shown in Fig. 1-7. Elements with 3 electrons in the valence band,

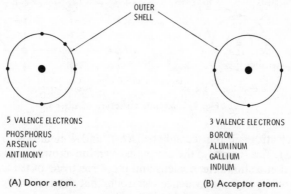

OUTER
SHELL

5 VALENCE ELECTRONS
PHOSPHORUS
ARSENIC
ANTIMONY

3 VALENCE ELECTRONS
BORON
ALUMINUM
GALLIUM
INDIUM

(A) Donor atom.

(B) Acceptor atom.

Fig. 1-7. Simplified drawings of impurity elements.

such as boron, aluminum, gallium, and indium, are called *acceptor* materials. Elements that have 5 valence electrons are phosphorous, arsenic, and antimony. They are called *donor* atoms. The doping of the silicon crystal determines whether it acts as a *p-type* or *n-type* material.

If the impurity that is added has 5 valence electrons, the bonding of the material is as shown in Fig. 1-8A. Note that at the position where there is an impurity atom, there is an extra electron in the outer orbit. Such a crystal is called n-type because doping with a donor material has produced extra electrons which can be moved as negatively

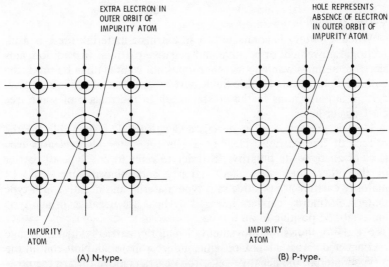

EXTRA ELECTRON IN
OUTER ORBIT OF
IMPURITY ATOM

HOLE REPRESENTS
ABSENCE OF ELECTRON
IN OUTER ORBIT OF
IMPURITY ATOM

IMPURITY
ATOM

IMPURITY
ATOM

(A) N-type.   (B) P-type.

Fig. 1-8. Silicon lattice structures showing presence of an impurity atom.

charged particles. It is important to note that the semiconductor material itself does not have an overall negative charge because the total number of electrons in the substance equals the total number of protons. However, it is said to have electron carriers that can be moved with the application of a suitable outside force.

If the doping element has 3 valence electrons, the bonding of the atoms is such that there are electron vacancies as shown in Fig. 1-8B. This vacant charge position, at which there would be an electron in a complete bond, is called a hole. It constitutes an electron absence and can be considered as a positive particle. There is always a tendency for an electron from a neighboring atom to move into the empty position or hole. In this case, there will be a hole left in the bonded makeup of the neighboring atom. As a result, there is a free motion of holes (or positive charges) throughout the material. Current that results from the random motion of holes is said to be supported by the motion of hole carriers.

The actual conductivity of the material depends on the amount of doping. The higher the doping, the greater is the number of electron or hole carriers. Current in a semiconductor is usually spoken of as movement of positive or negative carriers instead of hole or electron carriers. In n-type semiconductor material, current is the result of the motion of negative charges (electrons), while in p-type material, current results from the motion of positive charges (holes). These positive or negative charges can be made to drift in a given direction to produce an electrical current.

Carrier Motion

When a battery is connected to an extrinsic material, there is a directional movement of positive and negative carriers. In addition, carriers (sometimes wanted and other times not wanted) can be produced by heat, light, strong electric fields, and other forms of radiation. Also, certain imperfections in the crystal result in the release of some free electrons or holes.

Heat is present in all materials and as such is responsible for the release of free carriers. Heat in a semiconductor also releases carriers of an opposite polarity. In order to refer to one type of carrier or the other, they are called majority and minority carriers. The majority carriers are positive in p-type material and negative in n-type material. Minority carriers released by heat are negative in a p-type material and positive in an n-type.

Fig. 1-9A shows the movement of majority carriers when a voltage is connected across a block of semiconductor material. Note that in the n-type material, the negative (electron) carriers move toward the positive terminal of the battery. The n-type material also contains minority carrier holes, and these positive charges move toward the negative terminal of the battery, as shown in Fig. 1-9B. This minority carrier motion is the result of heat in the semiconductor material. The resultant increase of these carriers with a rise in temperature is an undesirable feature in semiconductor devices. In practice, the number of minority carriers is small as compared to the majority carriers.

In the p-type material, majority carriers are positive particles or holes. They move toward the negative terminal of the battery as shown in Fig. 1-9A. The minority carriers are electrons; these negative particles move toward the positive terminal of the battery (Fig. 1-9B).

THE PN JUNCTION

The pn junction of a semiconductor is an important part of most solid-state devices—diodes, bipolar transistors, field-effect transistors,

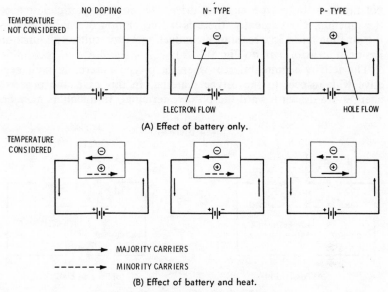

(A) Effect of battery only.

MAJORITY CARRIERS
MINORITY CARRIERS

(B) Effect of battery and heat.

Fig. 1-9. Charge flow in a semiconductor.

integrated circuits, and so forth. The types and applications of such junctions seem endless because of the versatile manner in which characteristics can be controlled by the shape, extent of doping, type of material, manner of activation, and other factors.

What occurs at a pn junction is best clarified by considering the junction activity—in terms of the motion of positive (hole) and negative (electron) charges. Unlike charges attract; like charges repel. Therefore, the electron is a negative charge that can be attracted by a positive voltage or charge but repelled by a negative voltage or charge. Conversely, the hole is a positive charge that can be attracted by a negative voltage or charge but repelled by a positive voltage or charge. Current in a semiconductor is composed of a movement of negative or positive charges and, in some cases, motion of both negative and positive charges in opposite directions.

Charge Motion

Much happens when a piece of n-type semiconductor is "positioned" back-to-back with a piece of p-type material as shown in Fig. 1-10. The motion of charges depends on the polarity of any external voltage applied across the junction. When a negative voltage is applied to the n-type material and a positive voltage to the p-type material, there is a motion of charges (current). The negative potential on the n-type material repels the negative charges, driving the electrons toward the junction between the two segments. In a similar manner, the positive

15

potential on the p-type material drives the positive particles toward the junction. Consequently, there is a free motion of charges across the junction and a low resistance-conducting path results. The junction has been *forward biased* (Fig. 1-10A).

The activity is quite different when the p-type material is made negative with respect to the n-type material. In this case, the negative charges are drawn toward the positive terminal. In a similar manner,

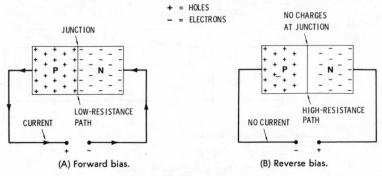

Fig. 1-10. The pn junction.

the positive charges are drawn toward the negative terminal. Therefore, charges are pulled away from the junction and there is no motion of charges between the two segments. Thus, a continuous charge motion (current) is not established because of the very high resistance of the junction under this bias condition. In this case, the junction is said to be *reverse biased* or back biased (Fig. 1-10B).

The pn junction just described has a low resistance when it is forward biased, permitting a high current. When it is reverse biased, it has a high resistance; little or no current results. External current can only be in one direction and, therefore, a pn junction can function as a diode detector or rectifier. This is the basic operation of a pn junction; however, there are other activities worthy of note.

Barrier Potential

When a pn junction is formed, the majority carriers near the junction attract each other (even without the application of a bias). They cross the junction and combine or cancel each other, as shown in Fig. 1-11A. This cancelling action of the carriers by electron-hole pairing establishes a charge between the two types of semiconductor material. Since the majority carriers near the junction have cancelled (combined), the semiconductor material near the junction has a charge that tends to hold the majority carriers away from the junction as shown in Fig. 1-11B. In effect, no more majority carriers can move to the junction because the electrons in the n-material are repelled by the negative charge of the p-material and the holes of the p-type material are repelled by the positive charge in the n-type material. The majority carriers, therefore, maintain positions back from the junction. This repelling force is called a barrier potential (Fig. 1-11C). In silicon, it amounts to a charge of approximately 0.5 volt. This barrier potential must be overcome before it is possible to move majority carriers (electrons or holes) freely across the junction.

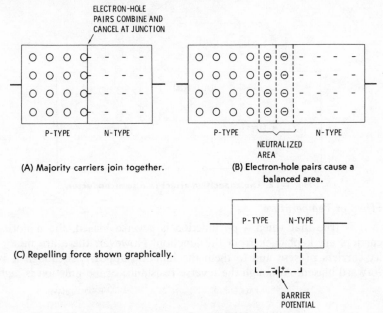

(A) Majority carriers join together.

(B) Electron-hole pairs cause a balanced area.

(C) Repelling force shown graphically.

Fig. 1-11. Forming the pn junction barrier.

One important thing about the pn junction is that the establishment of the barrier potential depends on the movement of electrons in one material and holes in the other. Electrons are bound, or immobile, in a p-type material; the current depends on the movement of holes. In an n-type material, the holes are immobile and the current depends on the movement of electrons. This fact is important to the operation of a pn junction and practically all semiconductor devices.

Junction Capacitive Effect

In the area near the junction where the limited combining of the majority carriers takes place, the resistivity is high. In fact, this section acts much like an insulating dielectric material between two plates of a capacitor. In the pn junction shown in Fig. 1-12, the edge of the majority carrier in each segment acts as a capacitor plate. When the junction is reverse biased by an external voltage, further cancellation of the majority carriers takes place by pairing, and the two demarcation lines move further back from the junction. In effect, a capacitance of a lower value is established. Thus, by changing the reverse bias value, one changes the effective capacitance of the junction. This is one manner of forming a capacitor in a monolithic integrated circuit.

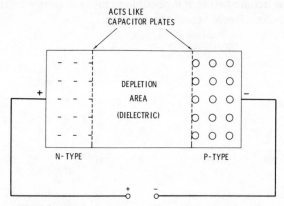

Fig. 1-12. The capacitive effect in a semiconductor.

Effect of Temperature

It is true that when a pn junction is reverse biased, the majority carriers are kept back from the junction. However, there are minority carriers present and to them the junction appears as though it is forward biased. Although the reverse resistance of the junction is high

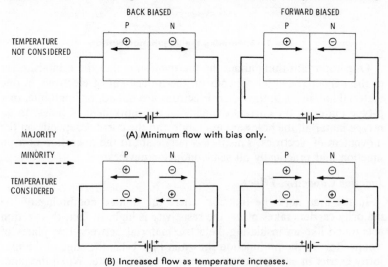

(A) Minimum flow with bias only.

(B) Increased flow as temperature increases.

Fig. 1-13. Minority carrier (reverse current) flow.

compared to the forward bias resistance, there will be a minority charge flow (Fig. 1-13). This current is low. At low operating temperatures, it is usually considered insignificant. However, at a high operating temperature, it is a factor that must be considered. Even when forward biased, a rise in temperature will increase the number of minority carriers and the external current increases correspondingly (Fig. 1-13B).

The response of a typical pn junction is shown in Fig. 1-14. After the forward bias voltage is made to exceed the barrier potential, the forward current begins its rise. The higher the forward bias voltage,

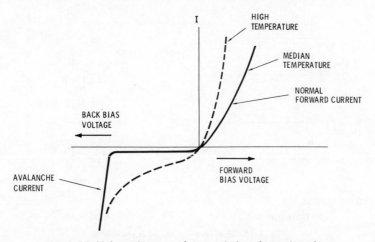

Fig. 1-14. Voltage/current characteristics of a pn junction.

the greater the junction current—up to normal operating limits. Current is low for the back bias condition and changes very little with an increase in the reverse voltage, up to the avalanche break-down value. At this potential, there is an abrupt increase in junction current. When the reverse bias exceeds a certain value, a complete breakdown of the electron binding occurs and the reverse current rises sharply. This is known as *avalanche current*.

The influence of temperature rise is shown by the dashed curve line. Note that for a given increase in forward bias, there is a higher junction current. With back bias, there is a current increase too. In effect, the efficiency of the pn junction as a unidirectional current device is decreased as a result of heat.

THE BIPOLAR TRANSISTOR

The most common integrated circuit device is the bipolar transistor. Scores of such transistors are often part of a single monolithic chip.

The bipolar transistor is a three-segment device having two pn junctions (Fig. 1-15). The center segment is called the base; the two outer segments are called the emitter and the collector. The base segment is a different polarity from the two outer segments, an n-type semiconductor for pnp transistors and a p-type semiconductor for npn transistors.

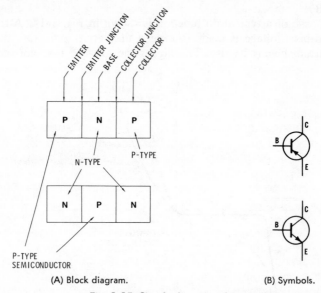

(A) Block diagram. (B) Symbols.

Fig. 1-15. Simple diagram of a transistor.

The point at which the emitter and base segments join is known as the emitter-base junction or simply the emitter junction. The position at which the collector and base join is called the collector-base junction or simply the collector junction.

The following factors are important in understanding the operation of a transistor:

1. In an operating circuit, the collector-base junction is reverse biased. Thus, the collector-base junction has a high resistance.
2. When no voltage (zero bias) is applied to the emitter junction, there is no current (charge motion) through either junction.
3. If the emitter junction is reverse biased, it has a high resistance. Again, there is no current through either junction.
4. When the emitter junction is forward biased, as it is in normal operation there is a motion of charges across the emitter junction. Its resistance is low. Internal activity is such that there is an amplified current crossing the collector junction. In a normal circuit, the collector current is higher than the base current. The tran-

sistor stage can be operated as a current amplifier and, with an appropriate external circuit, as a voltage amplifier.

Biasing

Consider the activity that takes place when in normal operation the collector junction is back biased and the emitter junction is forward biased. A small amount of signal current variation in the base circuit causes a substantially larger variation in the collector current. This collector current produces an output in the load circuit.

The circuits shown in Fig. 1-16 are biased for normal operation. In the case of the npn transistor, the collector junction is reverse biased by connecting a positive voltage to the collector, while the emitter junction is forward biased by connecting a positive voltage to the base and a negative voltage to the emitter. The forward bias on the emitter junction forces positive and negative charges to the junction, and a base current results. As a result, electrons from the emitter move into the base element. Some of them are neutralized by the positive charges or holes in the base. However, the emitter is doped to a higher level than the base. Consequently, more electrons move into the base than there are neutralizing holes. These electrons move toward the collector because of the positive potential of the collector. As a result, there is a strong movement of electrons from the emitter into the collector by way of the base. This current is greater than the base current.

A similar activity occurs in a pnp transistor. Pnp and npn transistors are said to have complementary characteristics. In the case of the pnp transistor, the forward biasing of the emitter junction causes a movement of positive charges across the emitter junction and into the base. These are, in part, neutralized by the electrons of the base. However, the positive charges continue on, attracted by the negative potential of the collector. Thus, they cross the collector junction in quantity and there is a strong collector current.

As shown in Fig. 1-16, current directions are opposite for the npn and pnp types. Note that a negative potential is applied to the p-type

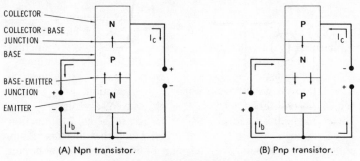

(A) Npn transistor.   (B) Pnp transistor.

Fig. 1-16. Basic transistor operation.

collector material in order to reverse bias the collector junction. Hence, the current direction in the external circuit is opposite from that of the npn transistor circuit.

Gain

In the previous discussion, the dc current paths were considered. The ratio of the dc collector current to the dc base current is referred to as the dc current gain or dc beta of a bipolar transistor. In the common-emitter configuration shown in Fig. 1-17, the dc current gain is represented by the symbol h_{FE}. Another current gain term is alpha, which is the ratio of the emitter current to the collector current. The value for alpha is always less than unity.

An audio signal applied to the base of the circuit shown in Fig. 1-17A causes the base current to vary about the base-bias current set by the dc base-bias voltage. This base-current variation, in turn, causes an amplified change in collector current. This can be seen by comparing the collector current variation with the base current variation of Fig. 1-16B. The ratio of the ac collector current to the ac base current is known as the small-signal beta of the transistor.

$$h_{fe} = \frac{\Delta I_c}{\Delta I_b}$$

where,

h_{fe} is the small-signal current gain,
ΔI_c is a small change in collector current,
ΔI_b is the corresponding small change in base current.

Output Voltage

The actual output voltage E_O results from the variations of the collector current in the collector load R_L. The input base-current variation

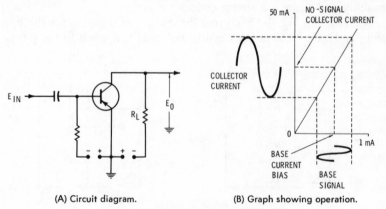

(A) Circuit diagram. (B) Graph showing operation.

Fig. 1-17. Basic transistor amplifier.

was caused by the input voltage E_{IN}. Hence, the actual voltage gain of the transistor stage is the ratio of E_O/E_{IN}.

In the basic common-emitter circuit, the input and output voltages are out-of-phase. In Fig. 1-17A, a pnp transistor is indicated; its base is negative with respect to the emitter, and its collector is negative with respect to the base. As shown in Fig. 1-18, a positive swing of the base voltage decreases the base current. In turn, the collector current decreases. A decrease in collector current through the load resistor makes the output voltage swing negative. Conversely, the negative al-

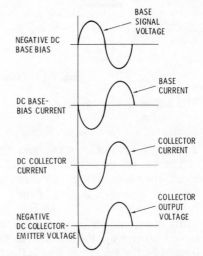

Fig. 1-18. Basic amplifier waveforms.

ternation of the input voltage causes an increase in base current, an increase in collector current, and a positive swing of the output voltage.

Transistor Currents

The direction of electron flow into and out of the two transistor types is shown in Fig. 1-19. Arrows indicate that the emitter current is a combination of the base and collector currents. This fact is apparent when you recall that most of the carriers emitted into the base move to the collector to produce the collector current, while a few of the carriers combine with opposite polarity carriers in the base to produce the base current. Stated as an equation:

$$I_E = I_B + I_C$$

The base current can be indicated as the difference between the emitter current and the collector current. By rearranging the formula:

$$I_B = I_E - I_C$$

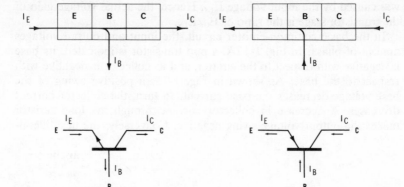

(A) Npn (B) Pnp

Fig. 1-19. Transistor currents.

The collector current is:

$$I_C = I_E - I_B$$

This applies to both the pnp and npn types shown in Fig. 1-19.

The matter of diffusion current in the base is often a stumbling block in understanding the operation of a transistor. Actually it is the result of the higher doping of the emitter segment as compared to the base segment. When the emitter junction is forward biased, a great many of the emitter carriers cross the junction into the base. As mentioned previously, the number of electron-hole pairs that combine is limited by the fewer carriers in the thin base. The bulk of the carriers crossing between emitter and base are not neutralized. They crowd each other, Fig. 1-19A. Since they are of like polarity, they repel each other and drift across the relatively thin base segment toward the collector segment. In effect, this mutual repulsion causes the carriers to diffuse to a region of low concentration (Fig. 1-19C) from the area of high concentration on the other side of the emitter junction. Once they

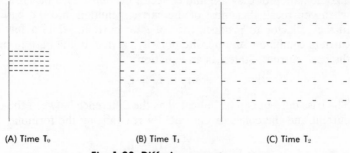

(A) Time T_0 (B) Time T_1 (C) Time T_2

Fig. 1-20. Diffusion current.

arrive at the collector junction, they are attracted across this demarcation by the opposite polarity of the collector. They are then moved at high velocity through the collector and into the external collector circuit.

The thinner the base, the shorter the time interval needed for the carriers to diffuse from the emitter side of the base to the collector side of the base. As the carriers reach the collector, they must be removed from the area of the junction by the voltage potential applied to the collector terminal.

The collector voltage need only be high enough to sweep the collector area near the junction free of current carriers. Increasing the collector voltage does increase the collector current slightly, but this increase is used primarily to obtain a more efficient removal of the carriers crossing the junction.

Remember that the collector current is the result of available carriers that are injected into the base by the forward biasing of the emitter junction. If the carriers are not injected into the base, they will not be available in the collector circuit and there will be no collector current. Thus, the collector current is related directly to the number of carriers injected into the base. This is fundamental for the operation of the bipolar transistor. Changing the emitter-junction bias influences the number of injected carriers and, therefore, varies the collector current. Fundamentally, a transistor is known as a current amplifier because the collector-current change responds to the change in the input current; namely, variations in the level of the injected carriers.

MAJORITY AND MINORITY CARRIER MOVEMENTS

The movements of majority and minority carriers for a pnp transistor are depicted in Fig. 1-21. The movement of positive particles (holes) is indicated by the (+) sign, while the negative sign (−) indicates the movement of negative particles or electrons. Fig. 1-21B illustrates the movement of majority carriers only.

With normal biasing, the electrons of the n-type base neutralize the holes injected into the base from the p-type emitter. For each electron that cancels an injected hole, there is an electron supplied to the base from the bias source. At the terminal end of the emitter, there must be an electron that leaves the emitter and enters the positive terminal of the base source. This current through the emitter junction and around the external circuit is called the *bias* or *base current*.

More holes are injected into the base from the emitter than can be neutralized by the base electrons, so these positive charges diffuse through the base to the collector. These holes present a positive charge at the collector junction and they are attracted to the negative potential of the collector. Electrons in the collector rush to pair up with

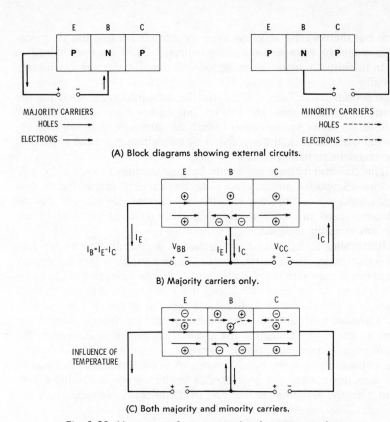

(A) Block diagrams showing external circuits.

MAJORITY CARRIERS

HOLES ⟶

ELECTRONS ⟶

MINORITY CARRIERS

HOLES ------►

ELECTRONS ------►

B) Majority carriers only.

$I_B = I_E - I_C$

(C) Both majority and minority carriers.

INFLUENCE OF
TEMPERATURE

Fig. 1-21. Movement of current carriers in a pnp transistor.

the incoming holes. As a result, at the terminal end of the collector, electrons must be supplied from the collector supply-voltage source.

As shown in Fig. 1-21B, the majority-current path is from the emitter to the collector through the two junctions. However, in the discussion of separate input (emitter-base) and output (collector-base) circuits, the currents are thought of as existing in the base lead as shown by the two arrows labeled I_E and I_C. These two currents are in opposite directions; consequently, the current present in the base lead is the difference between the emitter current and the collector current.

The majority-carrier movement (solid line) and the minority-carrier movement (dashed lines) are shown in Fig. 1-21C. In normal operation, the emitter junction is forward biased. Therefore, the minority carriers which are present (electrons in the p-type emitter and holes in the n-type base) move away from the junction. Thus, the minority current is insignificant and is usually not considered. Were the emitter junction reverse biased, there would be no majority current but there would be some minority current because minority carriers would be propelled toward the junction.

In the normal operation of a transistor, the collector junction is reverse biased. As a result, the minority carriers—electrons in the p-type collector and holes in the n-type base—are propelled toward the collector junction and there is a minority current. It should be noted that the majority carriers moving through the base, as a result of the carrier injection from the emitter, are of the same polarity as the minority carriers that are generated in the base by heat. These carriers are also neutralized by the electron carriers in the collector.

The minority current in the collector circuit when there is zero voltage present at the emitter junction is called the *collector cut-off current,* I_{CO}. With normal operation and biasing of the emitter junction, there is a resultant collector current which is the sum of the majority-carrier collector current and the minority-carrier collector current. (The latter is related to the effects of temperature.) Stated as an equation:

$$I_C = (I_E - I_B) + I_{CO}$$

One of the hazards of minority-carrier flow is a distinctive effect called *thermal runaway.* Heating can cause the minority current to forward bias the emitter junction. The direction of minority current is such that the emitter junction is forward biased even though no external forward bias is being applied. This produces a small collector current. As the junction is heated, there are more minority carriers formed. Thus, minority current increases and places a higher forward bias on the emitter junction. This results in an increased collector current and further heating of the junction. This chain reaction multiplies rapidly; the resulting high current and heating can destroy the transistor. Appropriate circuit design can reduce the possibility of thermal runaway.

Even in many circuits that do not runaway, heat causes higher transistor currents. Thus, heat influences the operating condition of a transistor even though the transistor itself does not take off. Correct circuit design is needed to minimize the influence of heat on the regular operating conditions of a bipolar transistor stage. Certainly this is an important consideration in the design of integrated circuits that use many individual bipolar transistors; each of which is small, closely related physically, and often closely related electrically to each other. Heat limits the power capability of the average monolithic integrated circuit. High-powered integrated circuits are usually of the hybrid type.

Except that it illustrates the movements of majority and minority carriers in an npn transistor, Fig. 1-22 is the same as Fig. 1-21. The previous discussion will explain the actions in Fig. 1-22, if care is taken to observe proper polarities.

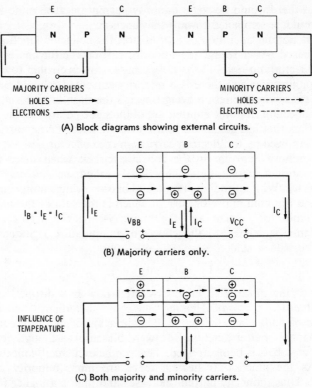

(A) Block diagrams showing external circuits.

$I_B = I_E = I_C$

(B) Majority carriers only.

INFLUENCE OF
TEMPERATURE

(C) Both majority and minority carriers.

Fig. 1-22. Movement of current carriers in an npn transistor.

TRANSISTOR FABRICATION

In the fabrication of modern transistors and integrated circuits, there are only two major processes although there are a number of additional techniques used less often. Diffusion and epitaxial processes are the most common and permit the construction of devices as shown

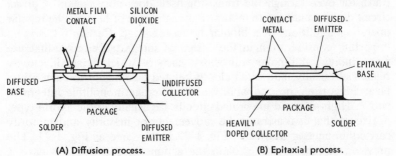

(A) Diffusion process.

(B) Epitaxial process.

Fig. 1-23. Methods of fabricating transistors.

in Fig. 1-23. In the manufacture of a bipolar transistor, there must be three separate semiconductor regions processed to form the collector, base, and emitter.

The construction process begins with a waferlike piece of semiconductor material called a substrate (one to two inches in diameter). Usually this wafer also serves as the collector of the finished transistors. In the diffusion method of construction, impurity atoms at high temperature are placed in contact with the surface of the substrate collector. This causes the impurity atoms to penetrate the material in a controlled fashion. In the epitaxial process, the additional segments are grown on top of the basic substrate collector. This buildup occurs atom by atom and includes the particular polarized impurity that is to form the next segment of the transistor. This is done in a heat chamber. Both processes are used in the fabrication of many transistor and integrated circuit types.

The transistor in Fig. 1-23A is called a double-diffused planar type. It begins with a silicon wafer about the size of a quarter or larger. This wafer is then diffused with the proper impurity, forming the substrate collector material. The remaining two elements of the transistor are then diffused into the wafer. Finally, the wafer is sliced into individual transistors; sometimes as many as 20,000 devices.

In the manufacture of an npn transistor, the basic silicon wafer is diffused with an n-type impurity atom. An insulating layer—usually silicon dioxide—is deposited on one side of the wafer. The other side must make an ohmic contact to provide a collector termination. This is often made to the case of the transistor after the wafer is sliced into individual transistors. Masking and photolithographic procedures are used to set up a design pattern for the diffusion of base and emitter impurities into the wafer. Usually the first diffusion step runs impurities into the central base area of the transistor. In our example, this would be a p-type impurity. In the second diffusion step, there is an n-type impurity run into selected areas of the wafer to form the emitter. Isolation between emitter and base surfaces is handled by a previously deposited insulating layer. A deposited metal film can be used to make ohmic contact between the base and emitter and the external transistor leads.

In the epitaxial process (Fig. 1-23B), the base and emitter segments are grown on top of the basic semiconductor wafer. This takes place in a high-temperature reaction chamber. The deposition becomes an extension of the crystal lattice of the wafer. The process begins with a lightly doped base grown onto the highly doped collector wafer. Material with opposite-type doping is now diffused into the base to form the emitter region. (The appropriate photolithographic and masking steps must accompany this single-impurity diffusion procedure that will complete the epitaxial transistor.)

FIELD-EFFECT TRANSISTOR (FET)

Another device inherent in a number of monolithic and hybrid integrated circuits is the field-effect transistor. There are two basic types, the junction field-effect transistor (JFET) and the insulated-gate type (IGFET). The field-effect transistor is called a unipolar device. It consists of a single junction and one polarity of charge carrier (either electrons or holes). In contrast, the bipolar transistor has two junctions, and both polarities of carriers (electrons and holes) are essential to its operation.

Junction FET

The junction field-effect transistor consists of a bar of p- or n-type silicon semiconductor material. A thin strip of semiconductor material with opposite charge carriers is placed above or wrapped around the bar (Fig. 1-24). In FET terminology, the bar is called a channel, while the element that controls the motion of charges along the channel is known as the gate. The common end of the channel is called its source, while the opposite end is named the drain.

An understanding of FET operation is aided by a review of semiconductor junction theory. When a junction is reverse biased, there is no significant motion of charges across the junction. However, as shown earlier, there is a rearrangement of the electron and hole carriers. Electron and hole carriers appear drawn away from the junction, establishing a *depletion area* (Fig. 1-25A). An increase in the reverse bias, as shown in Fig. 1-25B, widens the depletion area. The width of the depletion area in both the p- and n-segments also depends on the number of carriers within the area, and the amount of carriers is a function of the chemical makeup or doping of the semiconductor material. Since a greater number of excess charges are available per volume, the actual depletion area is smaller for a heavily doped material than for one with fewer charge carriers (Fig. 1-25C).

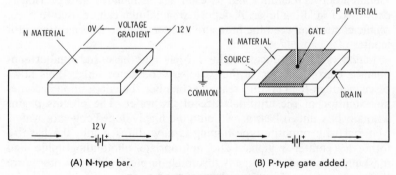

(A) N-type bar.

(B) P-type gate added.

Fig. 1-24. Basic FET structure.

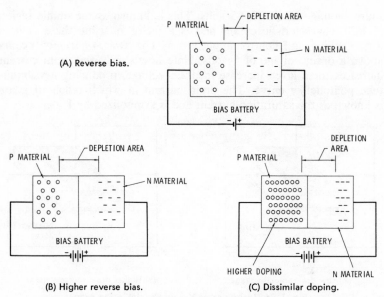

(B) Higher reverse bias. (C) Dissimilar doping.

Fig. 1-25. Influence of bias and doping on the depletion area.

A junction exists between the gate and the channel of the junction FET. A bias voltage applied to the gate is able to control the motion of charges (current) that flow between the source and the drain. When there is a zero bias between the gate and the source (see Fig. 1-24B), the application of a positive voltage between the drain and the source will cause the electron carriers of the channel to move between source and drain. As the positive voltage is increased, this current rises. In effect, the channel acts as a resistor.

As current rises, a *voltage gradient* (see Fig. 1-24A) appears along the semiconductor material (the resistor), and the junction between the gate and the channel becomes reverse biased. This causes a depletion area to extend outward from the gate into the channel, as shown in Fig. 1-26. This results in a decrease in the effective cross-sectional area of the channel, which is similar to decreasing the diameter of a conductor or a resistor. The greater the voltage potential gradient along the channel, the further the depletion area extends into the channel. Eventually, a point is reached at which there is no significant increase in drain current because the extension of the depletion area balances out the influence of any increase in drain-source voltage. The current is then said to be *pinched off*.

FET Characteristic Curves

The above explanation is represented by the $V_{GS} = 0$ curve shown in Fig. 1-27. Note how the drain current increases between a drain-source voltage of 0 and +5 volts. This is known as the ohmic region of FET characteristic curves and is a factor in using these devices for switching and control applications. The *pinch-off region* occurs above a drain voltage of 5 volts. Note how slowly the drain current increases for a given increase in drain voltage, producing a vacuum-tube pentodelike curve. The drain current at which pinch-off starts is known as the saturation current and is symbolized by I_{DSS}.

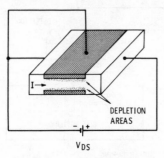 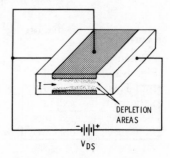

(A) Low drain-source voltage.   (B) Higher drain-source voltage.

Fig. 1-26. Influence of V_{DS} on the depletion area.

Also, observe in Fig. 1-27, that the drain-source voltage V_{DS} corresponding to the saturation current I_{DSS} is +5 volts. The bias voltage V_{GS} needed to cut off the drain current for this amount of drain-source voltage is known as the gate-source pinch-off voltage V_p. In the example, this falls between -1.8 and -2.0 volts V_{GS}.

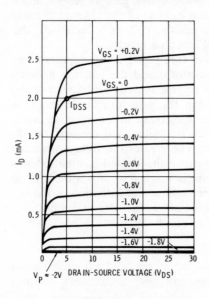

Fig. 1-27. FET characteristic curves.

32

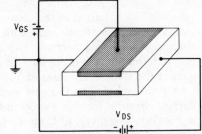

Fig. 1-28. Increasing negative bias on the gate.

When the gate is biased more negative in relation to the source (Fig. 1-28), the pinch-off point occurs at a lower value of drain current. This is to be anticipated because the negative voltage applied to the gate has extended the depletion area further into the channel. The higher the gate bias, the lower is the pinch-off drain current. Thus, individual bias curves can be plotted for the various values of gate bias, setting up a family of FET characteristic curves as shown in Fig. 1-27. Too much gate bias will close the channel and cut off the drain current just as a vacuum tube can be turned off by applying a high enough negative bias to its grid. In the example discussed, this value is about -2 volts V_{GS}.

FET Amplifiers

The FET has a very high input impedance as well as a moderate to high output impedance. In contrast, the bipolar transistor has a very low input impedance and an output impedance that is significantly lower than that of an FET. The FET is basically a voltage amplifier, while the bipolar transistor is a current amplifier.

A typical FET amplifier stage is shown in Fig. 1-29. An applied audio or other signal causes the gate voltage to vary about the gate

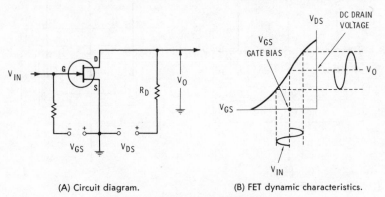

(A) Circuit diagram.   (B) FET dynamic characteristics.

Fig. 1-29. Basic FET amplifier stage.

bias set by the gate battery voltage. A small gate voltage variation causes a like variation in the channel or drain current. In fact, as the gate voltage is made to vary with signal, there results a substantial change in the drain current. The actual output voltage results from the variations of the drain current in the drain load resistor. Voltage gain of the stage is the ratio of V_O/V_{IN}.

A positive swing of the gate voltage produces an increase in drain current. In turn, there is a drop in the drain voltage (the output voltage swings negatively). Conversely, the negative alternation of the input voltage decreases drain current, and a positive swing of the drain and output voltage results. Input and output voltages are said to be of opposite polarity or out-of-phase.

The field-effect transistor is an excellent small-signal amplifier having low distortion and low noise. Like the pentode vacuum tube, it can be considered as a constant-current generator. Its transconductance, g_{fs}, is a measure of how much the drain current is changed by a given small change in gate voltage. In many practical applications, the equation for gain of a field-effect transistor is:

$$A_V - g_{fs}R_D$$

FET Fabrication

As with the bipolar transistor, diffusion and epitaxial methods are the common fabrication techniques. Instead of a single wrap-around gate element, a two-gate arrangement can be used. This technique provides two inputs for mixing and other circuit activities, such as agc or a feedback arrangement. These are known as double-gate field-effect transistors. In the basic junction FET, Fig. 1-30, a double diffusion process is employed. The basic semiconductor wafer serves

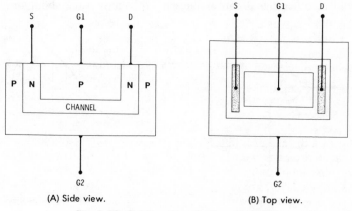

(A) Side view.   (B) Top view.

Fig. 1-30. Construction of a junction FET.

34

as the bottom gate region, and the drain-source channel is formed in the first diffusion process. At this time, the appropriate impurities are introduced through a mask. A second diffusion process is then used to form the top gate.

In the epitaxial method, it is possible to process a segment with two degrees of doping in order to establish the most favorable electrical characteristics near the junction. At the same time, this presents the most favorable conditions to the external circuit. In some cases, combined diffusion and epitaxial techniques are employed.

MOSFETs

A special form of the field-effect transistor is the MOSFET—metal-oxide semiconductor field-effect transistor. Its structure provides some unique characteristics and makes the MOSFET attractive for small-signal and low-noise, audio- and radio-frequency amplifiers. The characteristics are favorable for use in other high-impedance applications and in chopper and switching circuits.

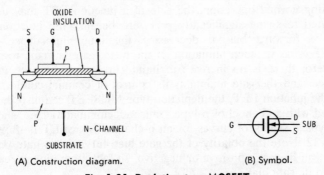

(A) Construction diagram.   (B) Symbol.

Fig. 1-31. Depletion-type MOSFET.

In this structural arrangement, illustrated in Fig. 1-31, the gate electrode acts as a control element just as it does in a junction field-effect transistor. However, in the junction type, the gate exerts an influence only when the junction is reverse biased. In the MOSFET construction, the thin gate layer is insulated by a thin layer of silicon dioxide, from the semiconductor material that forms the channel. The electrical field that results when the gate is biased appropriately influences the active carriers in the source-drain channel. Since there is insulation instead of a junction between the gate and the channel, the leakage current is very low and is little affected by temperature. A very high input impedance results. The gate and channel, separated by an oxide layer, are in effect the plates of a capacitor.

The arrangement of Fig. 1-31 shows the structure of a *depletion-type* MOSFET. In this example, the channel material is n-type so

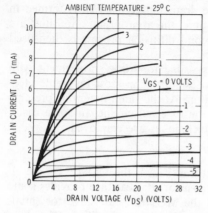

Fig. 1-32. Characteristic curves for a depletion-type MOSFET.

there are excess electrons. As the gate is made negative relative to the source, the number of electrons in the channel is depleted (reduced). This activity is similar to that of a junction FET.

During normal operation, the gate of a junction FET may not be permitted to swing significantly past zero because the junction will then be forward biased, decreasing the input impedance. The MOSFET has no such limitation. If the metal gate is made to swing past zero, there is no increase in input current or decrease in impedance, and the gate maintains its control of channel conductivity. Like the junction FET, the depletion-type MOSFET can also be constructed with a channel of p-type semiconductor material.

A typical family of curves for an n-channel MOSFET is shown in Fig. 1-32. Note the polarity of the gate bias curves. The gate voltage (V_{GS}) can be made positive or negative. In fact, operation is feasible with no dc gate bias.

Still another type of construction is the *enhancement-mode* MOSFET, shown in Fig. 1-33. Note that in its basic structure, there

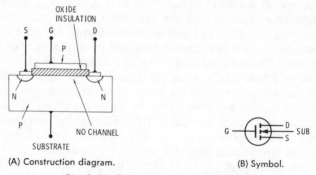

(A) Construction diagram.

(B) Symbol.

Fig. 1-33. Enhancement-mode MOSFET.

is no channel between the source and the drain. However, the gate itself does span this region. When a positive bias is applied to the gate, a conduction path between the source and the drain is established. The conductivity of this path is a function of the charge on the gate. In this manner, the magnitude of the drain current is controlled.

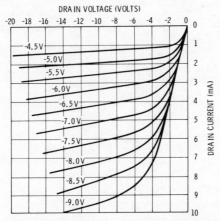

Fig. 1-34. Characteristic curves for an enhancement-type MOSFET.

The unusual characteristic of the enhancement-type MOSFET is that it is normally cut off (no drain current) until the application of a gate voltage. (The depletion-type MOSFET is the opposite—it is normally conductive. The application of a gate voltage causes the drain-source current to increase or decrease). When the gate-source junction of the enhancement MOSFET is at zero bias, there is no channel current and, therefore, no drain current. Note from the characteristic curves shown in Fig. 1-34, that drain current increases when the gate voltage is increased in the negative direction (away from zero). The higher the negative gate-source voltage, the higher the drain current. For ZERO gate bias, the drain current would be zero, while for −9 volts of bias (assuming a drain voltage of −12 volts), the drain current is nearly 10 milliamperes.

2

Integrated Circuit Structures

Integrated circuit fabrication is simply an extension, elaboration and miniaturization of the basic techniques discussed in Chapter 1. The transistor remains the key device and is formed in a similar manner. Processed simultaneously along with this basic device are diodes, resistors, capacitors and, occasionally, coils and other components. This chapter explains the various arrangements of integrated circuit components and how they are either made part of a single piece of silicon or incorporated into a hybrid design. Also covered are the various finished forms of integrated circuits and how they are connected into electronic systems. An experiment, based on several simple stages that can be built around an integrated circuit, concludes the chapter.

THE MONOLITHIC STRUCTURE

The conventional transistor is a three-layer affair that is usually built on a single substrate wafer which serves as the collector after the wafer is diced into individual transistors. The usual monolithic integrated circuit consists of two or more isolated transistors. Therefore, the continuous collector substrate is not feasible because it would not be possible to obtain isolation between the collectors of the individual transistors that comprise the integrated circuit. As a result, the basic monolithic integrated circuit is a four-layer affair that comprises three pn junctions, Fig. 2-1.

The substrate remains but it is no longer the collector layer. Instead, a second layer of opposite polarization is deposited on the substrate to serve as a collector. The individual collectors are now isolated from each other by the substrate material, as shown in Fig. 2-2.

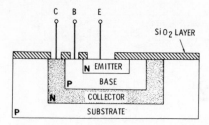

Fig. 2-1. A basic monolithic IC.

Additional diffusion and etching processes fabricate the base and emitter regions. Electrically, the junction between substrate and collector serves as an isolating diode. When this junction is back biased, each individual transistor can be isolated from all the other transistors and components of the integrated circuit. Then, dicing the wafer breaks up the fundamental wafer into individual integrated circuits, each having the appropriate number of transistors and other components that comprise a complete unit.

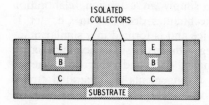

Fig. 2-2. Using substrate to isolate collectors.

In other integrated circuits, the isolation among transistors and other components is handled by a silicon-dioxide diffusion that serves as a dielectric (capacitive) isolator.

An electrically equivalent circuit for two transistors and the substrate of an integrated circuit is given in Fig. 2-3. Note that individual diodes link the collectors separately to the substrate. These diodes are formed by the pn junction between each collector and the common

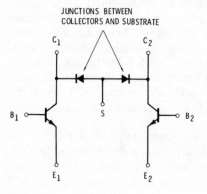

Fig. 2-3. Electrically equivalent circuit for the substrate and two transistors of an IC.

substrate. The collectors will remain isolated as long as these diodes are reverse biased.

Isolation is not complete because these junctions have a specific capacitance, as any pn junction does when it is back biased. This capacitance is a factor in the fabrication of integrated circuits and the use of such circuits in high-frequency applications. Fortunately, science has advanced greatly in the field of miniaturization, and low-capacitance high-frequency designs are common.

BASIC FABRICATION METHODS

The most common method of fabrication is known as the epitaxial-diffused method and begins with a p-type silicon wafer for the substrate (Fig. 2-4A). This is followed by an epitaxial n-type layer. Next, an isolating silicon-dioxide layer is added (Fig. 2-4B). Diffusion and masking processes are then used to complete the integrated circuit.

There are also two methods of fabrication that are strictly diffusion type. In the first, the substrate, again, is usually a p-type silicon wafer followed by a diffused layer of silicon dioxide. A phosphorous diffusion process now forms isolated collectors, as well as so-called n-type islands. The base and emitter elements are next diffused into these islands. In the other diffusion process, the substrate itself is n-type material and formed under a controlled-doping procedure. This is covered with an appropriate silicon-dioxide layer. The photolith process permits the formation of the n-type collector regions in the substrate. Two additional diffusions form bases and emitters.

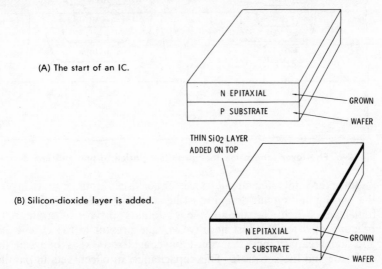

(A) The start of an IC.

(B) Silicon-dioxide layer is added.

Fig. 2-4. Fabrication of an integrated circuit.

Any one of the previous procedures adds an additional silicon layer. The composite consists of two devices, as shown in Fig. 2-5. The regular device is an npn type. However, the combination of base, collector, and substrate forms a parasitic pnp transistor as well. The collector-substrate junction must be kept back biased to keep the parasitic pnp transistor inactive.

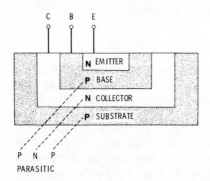

Fig. 2-5. The parasitic pnp transistor present in a basic monolithic structure.

In certain designs, it is possible to incorporate both types in a monolithic wafer if the electrical characteristics and isolation are not strict ones. A high order of electrical isolaton can be obtained using the five-layer arrangement of Fig. 2-6. In this arrangement, there are no elements common to both transistors.

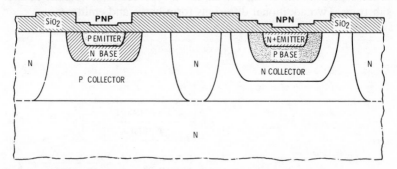

Fig. 2-6. Five-layer fabrication that permits isolation of pnp and npn devices.

Since the integrated circuit transistor consists of four or more layers, the biasing and circuit design must be such that it does not become a four-layer switch. To do so, the customary p-type substrate must be connected to the most negative voltage present in the circuit. An additional capacitance that must be considered exists between the collector and the substrate. This capacitance, in effect, acts in parallel with the normal base-collector capacitance of the transistor.

A TYPICAL MONOLITHIC PROCESS

The structure and some general electrical characteristics for a typical Motorola monolithic integrated circuit transistor are discussed in conjunction with Figs. 2-7, 2-8, and 2-9. The device is constructed using the epitaxial process and starting with a p-type substrate. Upon this is grown an n-type epitaxial layer that is 25 microns thick (1 mil). A thin silicon-dioxide film, resulting from exposing the wafer to an oxygen atmosphere heated to about 1000°C, follows.

The basic epitaxial wafer is now processed using a photolithographic process. This involves covering the silicon-dioxide layer with a photosensitive emulsion. A mask of the pattern to be fabricated is fitted upon the surface and the photolith exposed to ultraviolet light. The film not exposed to light is readily removed with certain liquids. The exposed portion forms a polymerized deposit over selected areas which cannot be etched. These remaining polymer areas now serve as a mask for the diffusion processes that follow.

As shown in Fig. 2-7, the diffusion activity penetrates between the highly resistant islands. By careful control, diffusion penetrates

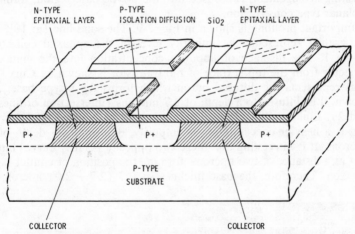

Fig. 2-7. The starting process in constructing an IC transistor.

through the n-type epitaxial layer down to the p-type substrate. Consequently, isolated n-type epitaxial islands are formed and are separated electrically from each other. A two-layer structure has now been formed that is topped with silicon-dioxide islands. The substrate and collector have been formed.

After the isolation step, two successive diffusions take place. First, with appropriate masking, the p-type base is diffused into the n-type collector (Fig. 2-8). Finally, the n-type emitter, through another

43

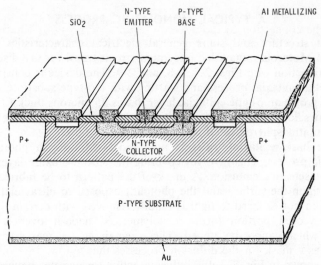

Fig. 2-8. Single isolated transistor of a monolithic integrated circuit.

masking arrangement, is diffused into the p-type base region, forming a planar type construction.

Important profiling is shown in Fig. 2-9. The solid line (at 10^{16} on the graph) represents the impurity level of the epitaxial collector, Curve 1. The distance of impurity concentration into the epitaxial collector from the top surface of the transistor is shown by Curve 2. This represents the diffused boron which forms the p-type base. As shown, it is diffused to a depth of 2.7 microns (intersection of lines 1 and 2). Emitter diffusion using phosphorous is depicted by Curve 3 and is a heavier concentration of impurity, diffusing to a depth of 2 microns. It is interesting that these two opposite polarity curves intersect at a distance of two microns, forming the position of emitter-base junction. Therefore, the base thickness is 0.7 (2.7 − 2.0) micron.

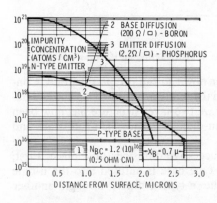

Fig. 2-9. Impurity profile of an IC transistor.

Making the connections is the final matter for consideration. Actually, the connection to the collector region can be provided simultaneously with the diffusion of the emitter by diffusing n-type impurities into the collector. Hence, an ohmic contact is made. An aluminum metalizing procedure that makes contact with each of the three elements is the final operation.

A comparison between a discrete planar epitaxial transistor and a single transistor of a monolithic integrated circuit is shown in Fig. 2-10. In the integrated circuit, an ohmic contact must be made to the collector which is isolated within the p-type substrate. This introduces an additional collector series resistance. As mentioned previously, there is also the additional capacitance that is present between the collector and substrate which does not exist for the discrete transistor.

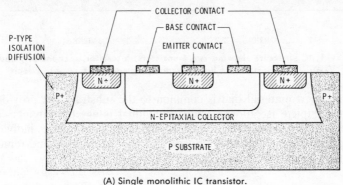

(A) Single monolithic IC transistor.

(B) Discrete bipolar transistor.

Fig. 2-10. A comparison of two transistors.

A multiphase structure is shown in Fig. 2-11. As mentioned previously for a basic monolithic arrangement, there is an orderly array of single-crystal silicon. Back biasing of the junctions provides an isolation for a given monolithic circuit. When individual monolithic groups must be isolated from each other, it is possible to use an elec-

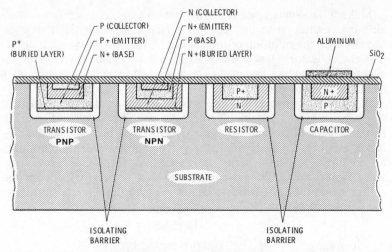

Fig. 2-11. IC structure showing two transistors, a resistor, a capacitor, and the isolating barrier separating each unit.

trically inert material that is common to the substrate but provides a more complete isolation between individual integrated circuits. This is called an isolating barrier. Such a barrier not only isolates individual integrated circuits but also reduces parasitic capacitance and removes the four-layer switching hazard.

ACTIVE COMPONENTS

A variety of active devices can be incorporated in an integrated circuit. The two most common active components of the modern integrated circuit are the transistor and the diode. The latter is most often a transistor connected as a diode. In addition to the bipolar transistor previously discussed, integrated circuits can include FET, MOSFET, and unijunction types. Besides the conventional diode, special types, such as the voltage-variable capacitance diode, tunnel diode, etc., also lend themselves to monolithic fabrication. Even such power devices as four-layer switches and silicon-controlled rectifiers are found in special type ICs.

Diodes

Although specific diodes can be fabricated within the IC chip, many integrated-circuit diodes are basically transistors with one element either left free or connected to another element. Transistors must be formed in the fabrication process, regardless of whether or not other devices are needed. This technique is inexpensive in comparison to the cost of incorporating special diode arrangements. The arrangement used depends upon the application of the diode in the IC circuitry.

The five ways in which a transistor can be connected as a diode are shown in Fig. 2-12. Each connection has its own characteristics. The most common connection is that of circuit A with the collector-base junction shorted. This wiring method takes out the parasitic pnp transistor. The connection provides good high-frequency performance, a high-speed diode operation in digital circuits, and the lowest forward voltage drop. When forward biased, the current is the function of

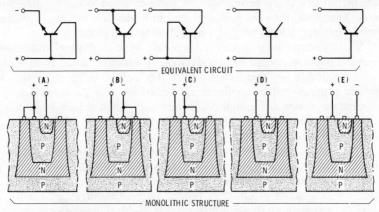

Fig. 2-12. IC transistors connected to operate as diodes.

minority-carrier flow and rises linearly with an increase in forward bias. The monolithic structure is shown in Fig. 2-13. The base and collector are tied together to serve as the diode anode; the emitter connection serves as the cathode. The substrate, as always, must be returned to the most negative point of the circuit, back biasing the collector substrate junction.

It should be noted that in all the other connections of Fig. 2-12, the collector and base are not shorted together. Therefore, the influence

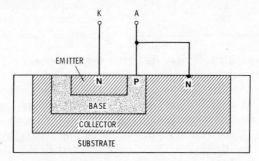

Fig. 2-13. Monolithic structure of an IC diode.

of the parasitic pnp transistor (substrate-collector-base) must be considered in terms of performance and circuit interconnections. Circuits B and D are attractive for switching circuits because of their ability to store a charge, particularly the floating collector of circuit D. The connections of circuits C and E serve well as general-purpose diodes and have a high reverse-voltage rating.

Variable-capacitance diodes in the form of varactor, parametric and tuning diodes can be incorporated in monolithic integrated circuits. However, tunnel diodes are made with much more favorable characteristics using germanium or gallium arsenite rather than silicon. Therefore, they are seldom found in monolithic structures. When used, they are usually part of a hybrid integrated circuit.

Variable-capacitive diodes incorporate special doping and structures. They are reverse-biased devices. A varactor diode is designed with a suitable nonlinear capacitance that serves as an efficient harmonic generator. The nonlinear characteristic of a parametric diode is capable of producing high-frequency gain when driven by a pump signal. The characteristics of the tuning diode are set by the rate and range of frequency change it must accomplish.

In all of the variable-capacitance diode functions, size and processing are important. Fine variable-capacitance diodes can be made in a silicon structure, although a few characteristics can be enhanced with the use of gallium arsenite. A typical monolithic arrangement for a variable-capacitance diode is given in Fig. 2-14. These devices require

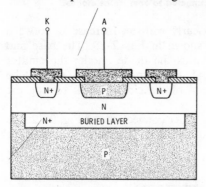

Fig. 2-14. Diagram of an IC variable-capacitance diode.

a low series resistance. Note that both the anode and cathode ohmic contacts are made from the top, as is required for the monolithic-strata type of construction. To lower the cathode-region resistance, a so-called "buried layer" of very high doping provides a low-resistance internal lateral-current path (high-plus doping) through the cathode structure. At the junction itself, the n-material is less highly doped, establishing the desired pn junction characteristic for the variable-capacitance diode.

Switches and Rectifiers

Normally, pnpn switches and silicon-controlled rectifiers are four-layer devices that consist of npn and pnp transistors, sharing a common pn control junction (Fig. 2-15A). This is a structure that is inherent in integrated-circuit construction, consisting of the npn transis-

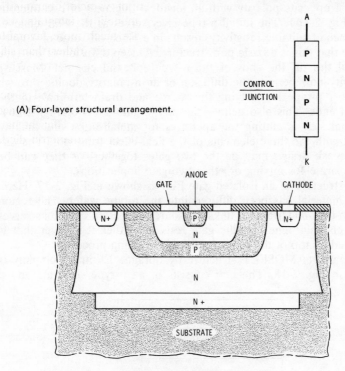

(A) Four-layer structural arrangement.

(B) Diagram of a five-layer devices.

Fig. 2-15. Multiple layer devices.

tor and its parasitic pnp neighbor. However, control geometry and the need for top contacts are such that a more suitable construction can be arranged using five layers, as shown in Fig. 2-15B. The p-segment at the top of the IC serves as the anode; the gate area is located between it and the cathode. Internal resistance is again reduced with the use of the highly doped n+ buried layer. In the operation of these devices, the common junction is reverse biased and there is no conduction. However, upon application of a *high reverse bias,* the junction avalanches and there is a low-resistance high-current path between the cathode and anode. It is the gating action at the common junction that switches the rectifier action of the device.

FETs

The three types of FET devices lend themselves well to monolithic IC fabrication. An example of the makeup of an epitaxial-diffused junction FET is shown in Fig. 2-16. The p-type substrate, that also serves as gate 1, is the basic wafer. Upon it are grown two epitaxial layers of opposite polarity with an identical thickness of two microns (A of Fig. 2-16). The initial n-type layer serves as the FET channel and, when of suitable resistivity, results in a pinch-off value of 4 volts.

Note that this is a basic pnp structure. A deep diffusion of p+-type material ties off the ends of the n-type epitaxial channel. Masking selectivity now permits the diffusion of an n-polarity doping through the upper p-type layer, tying the source and drain terminals to the channel ends. This also defines the p-type gate 2 (B of Fig. 2-16). The final step is cutting the apertures for metalization and making contact with the three elements of the field-effect transistor. The diffusion mask design may tie the two gates together, or they can be kept separate for mixing or other two-gate applications.

The structure of an isolated-gate FET is shown in Fig. 2-17. Here, n-type material has been diffused into the p-type wafer. The region of the p-type material that links the source and drain diffusions serves as the channel. The metallic gate rests on a dielectric layer that is deposited on top of the wafer by using a masking process.

A dual-gate MOSFET structure by Sprague Electric Company is shown in Fig. 2-18. The unit consists of an n-type substrate and a

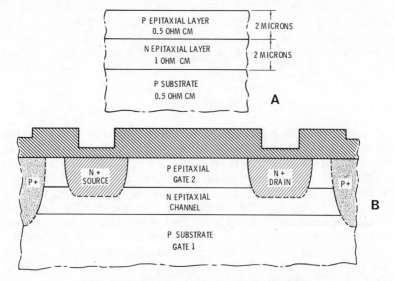

Fig. 2-16. An epitaxial-diffused junction FET.

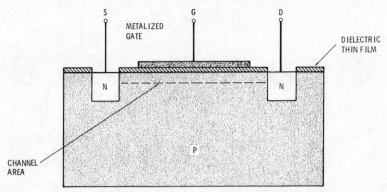

Fig. 2-17. Structural diagram of an IC IGFET.

p-type channel (with its source and drain connections). Two separate gates exert a balanced influence on channel operation through a special *ion-implanted* region. The resistance and other characteristics of the region can be controlled exactly. In the implantation process, boron ions are permitted to penetrate the more conventional oxide dielectric region. These charged ions permit a smaller area to have a high and exactly controlled resistivity.

PASSIVE COMPONENTS

The single most common passive component is the resistor. In fact, most integrated circuits are a combination of transistors, diodes, and resistors. The second most common passive component, but not nearly so common as the resistor, is the capacitor. Occasionally, coils and filters are incorporated into integrated-circuit design, but these are usually found only in hybrid-type ICs.

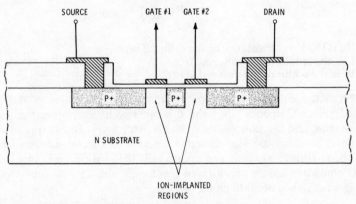

Fig. 2-18. Dual-gate MOSFET fabrication using Sprague Electric Company ion implantation techniques.

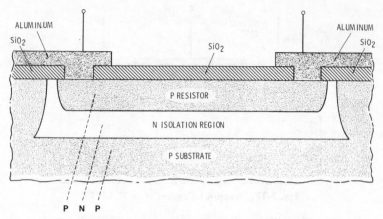

Fig. 2-19. Cross-sectional view of a diffused IC resistor.

In a monolithic integrated circuit, the fabrication of a resistor differs little from that of a transistor. Such a resistor in no way resembles the usual discrete resistor. It is, in fact, a highly doped silicon layer. A cross-sectional view of a diffused resistor is shown in Fig. 2-19. As usual, the process begins with the substrate. The resistors are formed simultaneously with the formation of the transistors and other segments of the monolithic integrated circuit. A n-type isolation region follows the substrate. This layer corresponds with the formation of the collectors in the neighboring transistors. Diffusion, along with appropriate masking, forms the layer of semiconductor material that serves as the resistor. This step is similar to the formation of the transistor base of a monolithic transistor. Contact leads are attached at opposite ends of this resistor layer. The equation for such a resistor is:

$$R = R_S\left(\frac{L}{W}\right)$$

where,

R_S is the sheet resistance of the diffused layer,
L is the length of the diffused area,
W is the width of the diffused area.

From the equation it is apparent that the ohmic value of the diffused resistor depends upon the doping, which determines the sheet resistance, and the dimensions of the resistor layer. In the tiny monolithic chip, there are size limitations and also sheet-resistance limits of about 10 to 500 ohms per square mil. If the sheet resistance were 200 ohms per square mil, a layer 1 mil wide and 25 mils long would have a resistance of 5000 ohms, or:

$$R = 200 \text{ ohms} \times \frac{25}{1} = 5000 \text{ ohms}$$

A practical range of values for monolithic diffused resistors extends between 50 ohms and 30,000 ohms.

It should be noted from Fig. 2-19 that the resistor itself is part of the parasitic pnp transistor and includes its stray capacitances. Therefore, it is necessary to connect the substrate to the most negative voltage present in the circuit, keeping the parasitic transistor isolated and incapable of four-layer switching. It is also important that the pn junction between the resistor and the n-type isolation layer never becomes forward biased. To prevent this, it is necessary that the isolation layer be held at the highest positive circuit potential.

Very tiny resistor chips can also be fabricated for use in hybrid integrated circuits. The arrangement shown in Fig. 2-20 is typical.

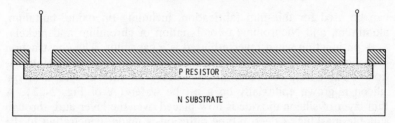

Fig. 2-20. Fabrication of resistor chip for use in a hybrid IC.

The n-type substrate is basic and provides the mechanical strength, while the p-type layer is the resistive element. Such a resistor is well isolated and need only contend with the pn diode junction and its distributed capacitance when a contact is made to the substrate material.

Resistors are subject to frequency effects, monolithics more than hybrids, because of the inherent capacitance between the resistor and the isolating substrate. Therefore, the impedance of the combination—resistance and stray capacitances—drops off at the higher frequencies. Resistors of this type have a positive temperature coefficient; that is, resistance increases with temperature.

The ion-implantation technique (Fig. 2-21), permits the processing of very shallow pn-junction resistors. In the example, the implantation is made into the n-type epitaxial material using appropriate photomask techniques. The penetration of the charge ions can be controlled precisely and resistor values of high accuracy can be obtained. Sheet resistivities can be varied between 500 and 10,000 ohms/square mil.

The thin-film technique can also be used to form integrated-circuit resistors. In hybrid circuits, such a thin film of metal or semiconducting oxide is laid down on a substrate. In monolithic application, the thin film can be deposited on the regular silicon dioxide layer using mask etching. An insulating layer is now placed on top of the resistor and apertures etched for making ohmic contacts. Various materials

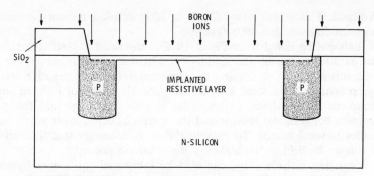

Fig. 2-21. Using the ion-implantation process to fabricate a resistor.

can be used for thin-film fabrication, including tin oxide, tantalum, aluminum, and Nichrome (a combination of chromium and nickel).

A pn junction when reverse biased is a capacitor. The construction of a monolithic capacitor follows along the same lines as the fabrication of active transistors and diodes. First, a 25-micron layer of n-type silicon is grown epitaxially on a p-type wafer (A of Fig. 2-22). A thin layer of silicon dioxide is now placed over the layer and, through suitable masking, a deep p-type diffusion is made, penetrating to the

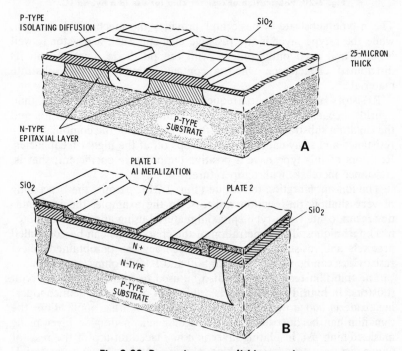

Fig. 2-22. Processing a monolithic capacitor.

substrate. This sets off n-type islands, serving as one plate of a junction capacitor. A shallow p-type diffusion now forms the second plate of the capacitor (B of Fig. 2-22). Ohmic contact is made to this area and to the n-type epitaxial layer to form the capacitor terminals.

Such a junction only operates as a capacitor when it is back biased properly. Also, the substrate must again be returned to the highest negative potential of the circuit to prevent conduction of the junction between the epitaxial layer and the substrate.

The construction of a tiny chip capacitor for hybrid IC application is shown in Fig. 2-23. Since such components can be isolated from one another on the ceramic wafer of a hybrid integrated circuit, they

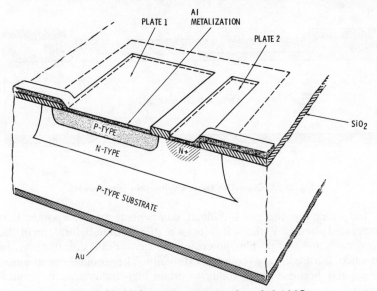

Fig. 2-23. Makeup of a capacitor for a hybrid IC.

are easier to fabricate and relatively free of parasitic influences. The capacitor is the reverse-bias junction between the p region and the n+-diffusion. External connections are made to the metalized aluminum areas on the top.

The thin-film process can also be used to form capacitors. One such device is the silicon dioxide capacitor illustrated in Fig. 2-24. In this case, silicon dioxide serves as the capacitor dielectric while the two capacitor plates are the aluminum metalization on top and the silicon substrate beneath. The substrate makes ohmic contact with the gold metalization at the very bottom. This idea can be included as part of a monolithic integrated circuit also, although additional diffusions are required—as in the case of the monolithic pn-junction capacitor. Other dielectric materials that can be used in the construction of thin-film type capacitors are tantalum oxide, aluminum oxide, and silicon monoxide.

Obtainable capacitor values are somewhat limited. In the case of monolithic construction, the maximum value that can be obtained is about 500 pF. For the hybrid types, a maximum value up to about 0.1 μF can be expected. Thin-film values go up to about 1000 pF. Larger value capacitors are usually in the form of separate packages for hybrid applications. They are generally in the form of a nonsilicon construction, such as a tantalum capacitor mounted in a hermetically sealed package or a TO-5 case.

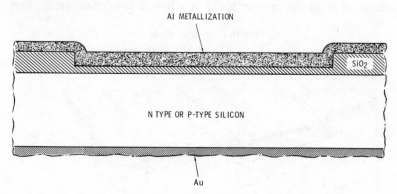

Fig. 2-24. Capacitor formed using thin-film process.

Inductors are the most difficult components to incorporate into integrated circuits. Perhaps it is just a matter of time before one of the semiconductor or thin-film processes will permit coil construction. In practice, most coils are mounted externally. The toroidal construction is popular because of its ability to attain high inductance in a small space.

It is feasible to incorporate various filters using resistors and capacitors into an integrated circuit. Those that require the presence of a coil are much more difficult to make.

PACKAGES AND SOCKETS

Integrated circuits come in a variety of cases and mounting arrangements. There are three major cases. One of these (Fig. 2-25) is the TO-5 round package with either 8, 10, or 12 leads. Another is the flat pack with 14 terminals, as shown in Fig. 2-26. This type of integrated circuit is usually surface supported and is wired directly into the printed-circuit board.

The most common types are the 14-terminal and 16-terminal dual-in-line cases (Fig. 2-27). These can be wired directly to a printed-circuit board having the appropriate aperture spacing or, as has become increasingly popular, mounted using 14-pin and 16-pin dual-in-line sockets. The dual-in-line cases are either plastic or ceramic packages. The ceramic packages, of course, are able to withstand higher temperatures. Also, TO-5 sockets are now widely used for that particular case style.

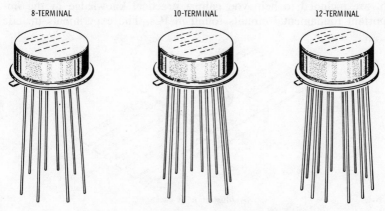

8-TERMINAL   10-TERMINAL   12-TERMINAL

(A) Round package having 8-, 10-, or 12-terminal leads.

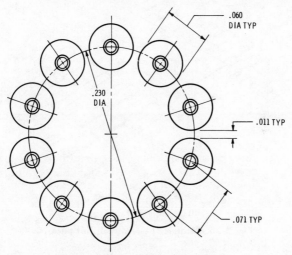

(B) Dimensions of the 10-terminal IC package.

Fig. 2-25. TO-5 package styles.

The dual-in-line type can be either surface mounted or have the leads forced through the holes of the pc board. Dual-in-line cases are found with as few as 8 and as many as 16 terminals. There are also staggered in-line cases as shown in Fig. 2-28. This style is used for high-current application and when greater separation between terminals is required or is desired (in terms of circuit board wiring).

EXPERIMENTS

A set of five experiments, one each at the end of Chapters 2 through 6, are included to help you gain a practical knowledge of the important fundamental circuits found in ICs. The experiments include

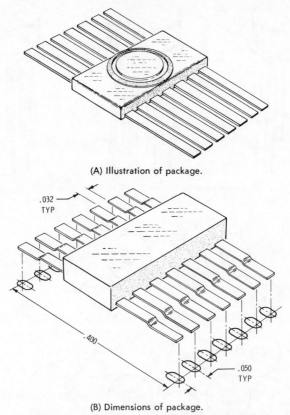

(A) Illustration of package.

(B) Dimensions of package.

Fig. 2-26. Ceramic, 14-terminal, flat pack IC.

the construction and tests of a differential amplifier, an operational amplifier, a crystal-controlled pulse generator, and a NOR-gate flip-flop and decade divider.

All five experiments are to be constructed on a single vector board (Fig. 2-29). TO-5 and in-line sockets are to be attached to the perforated vector board. These sockets are widely used in both experimental and finished IC circuits.

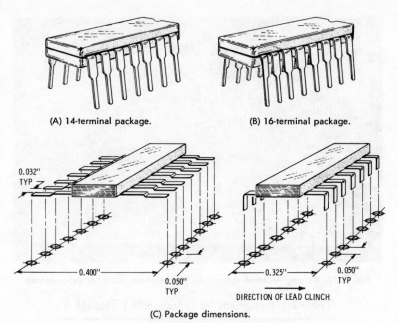

(A) 14-terminal package.

(B) 16-terminal package.

0.032"
TYP

0.400"

0.050"
TYP

0.325"

0.050"
TYP

DIRECTION OF LEAD CLINCH

(C) Package dimensions.

Fig. 2-27. Popular dual-in-line plastic packages.

The parts list shown in Table 2-1 includes the components required for all five experiments. Cost of the components is low, and you will gain practical experience as well as firming your theoretical knowledge of IC operation by constructing these projects.

In the first experiment, you will work with a simple four-transistor integrated circuit. You will check the performance of the associated transistors in both dc and ac applications. The final segment of the

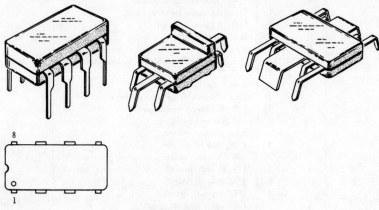

8

1

Fig. 2-28. Other IC cases.

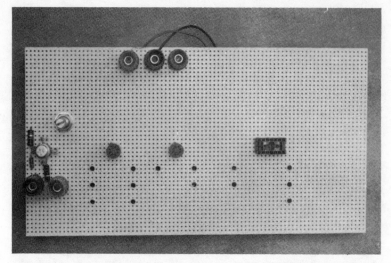

Fig. 2-29. Arrangement of sockets on vector board used for experiments.

Table 2-1. Equipment for Experiments 1 Through 5

1	Volt-ohm-milliammeter
1	Audio sine/square wave oscillator
1	Oscilloscope
1	Voltmeter, dc
1	100-kHz crystal and socket
1	Vector board, 8½″ × 4½″
2	HEP 580 IC
1	HEP 583 IC
1	741 operational amplifier
1	7490 decade divider
3	8-pin round IC socket
1	15-pin dual-in-line IC socket
1	Transistor radio battery, 9 volt
2	Lantern battery, 6-volt
2	Penlight battery, 1½-volt
12	Binding post
24	Vector micro-clip terminals (Vector Company)
1	7-35 pF trimmer capacitor
8	0.1-μF disc capacitor
1	2-μF 12-volt electrolytic capacitor
1	220-ohm ½-watt resistor
3	1K, ½-watt resistor
1	1.8K, ½-watt resistor
2	4.7K, ½-watt resistor
3	10K, ½-watt resistor
2	15K, ½-watt resistor
2	27K, ½-watt resistor
3	100K, ½-watt resistor
3	1-megohm ½-watt resistor
	Insulated hookup wire

experiment is the construction of a sine-wave audio oscillator. This audio oscillator can be used as a signal source in some of the succeeding experiments.

EXPERIMENT 1: FOUR-TRANSISTOR IC

General

The schematic diagram of a Motorola HEP 580 integrated circuit is given in Fig. 2-30. It consists of four transistors and six resistors. The transistors are connected in pairs with common collectors. The base inputs are separate for ease of connection in a differential amplifier circuit. All emitters are joined together and connected to pin 4. An external high-value resistor or constant-current source can be con-

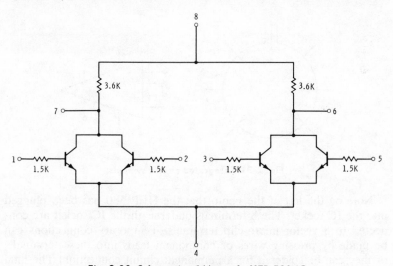

Fig. 2-30. Schematic of Motorola HEP 580 IC.

nected here for stabilizing a differential amplifier. Also, the emitter can be connected to a common or low-value emitter resistor for a more routine operation of the transistors. The proper value emitter-resistor and an appropriate feedback circuit permit the HEP 580 to be operated as an oscillator. Integrated series base resistances increase input resistances as well as suppress any tendency to parasitic oscillations.

Symbolically, integrated circuits are presented in the two ways shown in Fig. 2-31. The triangle representation is the more common, although the circle arrangement depicts a definite pin positioning. An advantage of the triangle representation is that schematic lay-outs can be drawn with greater ease, indicating input and output points regardless of pin numbers.

Vector Board Construction

The first step involves vector board layout. Except for the decade divider which will be the last experiment, the integrated-circuit sockets used are round 8-pin types. The first step, then, can be the mounting of four IC sockets on the vector board, three 8-pin round sockets and one 14-terminal dual-in-line socket. These can be seen in Fig. 2-29, mounted across the center of the 4½″ by 8½″ vector board. Binding posts and various other components for the first experiment are also shown in the photograph. (When the experiments are completed, the vector board will have mounted on it a tone oscillator, a square-wave 100-kHz crystal oscillator, and a frequency calibrator that has selected outputs from 100 kHz to 5 kHz.)

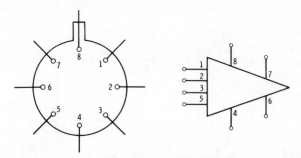

Fig. 2-31. Integrated circuit symbols.

Note on the left of the photo that the HEP 580 has been plugged into the IC socket. The 8 terminals underneath the IC socket are connected to 8 vector micro-clip terminals. Temporary connections can be made by pressing wires or component leads into these terminals, or they can be soldered for a permanent circuit connection. The final procedure of this experiment is the construction of a sine-wave oscillator. You may wish to wire this experiment permanently by making soldered connections. Supply voltage and oscillator output binding posts are also seen in the photograph.

Procedure 1: Basic Common-Emitter Stage

1. Connect the circuit of Fig. 2-32. Leave resistor R1 disconnected for the initial part of the experiment. Connect a current meter into the circuit and set to the 0-5-milliampere scale. Connect a dc voltmeter between the output terminal and common (ground).
2. Turn on the amplifier. Note that only a small leakage current is present. Supply voltages at pin 7 and pin 8 are approximately the same.
3. Connect a 1½-volt battery directly between the parallel connection

of pins 1 and 2 and common. The + voltage of the battery on the base input turns on the transistor. Note the current reading. Also, the collector voltage at the output drops to near zero. Is it reasonable to expect this drop in collector voltage? Remember the current is present in the internal 3.6K collector resistor. What will be the voltage drop across this resistor?

4. Reverse the base-bias battery voltage. Cutoff condition is established. Disconnect the battery and connect the 15K bias resistor (R1) into the circuit.

5. After a shut-down period of at least five minutes, turn on the amplifier. Record the collector current and collector voltage. In our test example, the collector current was 2.2 mA, and the collector voltage 3.1 volt. This reading will probably vary significantly from unit to unit.

6. Shunt a 10K resistor across the bias resistor R1. What happens to the collector current and collector voltage? Why?

A decrease in the ohmic value of the bias resistance increases the base bias current. As a result, collector current increases and collector voltage declines.

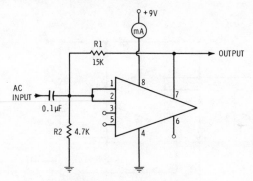

(A) HEP 580 and external circuit.

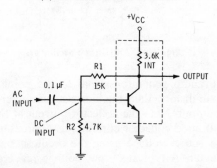

(B) External circuit and one internal transistor and resistor.

Fig. 2-32. Common-emitter amplifier.

Procedure 2: AC Operation

1. Substitute the 50K potentiometer for resistor R1. Connect the voltmeter to the output and the current meter to the supply voltage line. Turn on the amplifier.
2. Vary the potentiometer throughout its range. As the bias resistance is decreased, the collector current rises and the collector voltage falls. Adjust the potentiometer for a collector voltage of 4 volts. Note the collector current at this setting.
3. Apply a 1000-Hz sine wave to the input of the amplifier through the 0.1 μF series capacitor. Connect an oscilloscope to the amplifier output. Set the bias control resistor (potentiometer) to mid-position. Set the gain control of the audio generator to the minimum setting.
4. Slowly increase the audio generator output and observe the increasing amplitude of the sine wave displayed on the oscilloscope screen. Adjust the horizontal frequency control of the oscilloscope to display two sine waves.

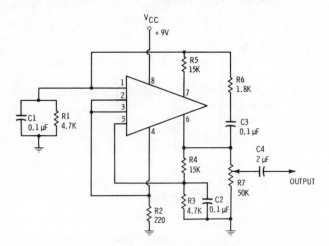

Fig. 2-33. An IC tone oscillator.

5. Increase the audio generator signal to the point where the sine wave begins to distort. Vary the bias potentiometer for a maximum undistorted output. Note the collector current at this point. In our sample, the desired collector current was approximately 1.8 mA.
6. Measure the output voltage using the oscilloscope or the ac scale of the voltmeter. A typical output voltage reading was 2.5 volts. Voltage gain of the amplifier is quite high and may be well over 50.
7. Disassemble the circuit.

Procedure 3: Tone Oscillator

1. Construct the tone oscillator using the simple feedback circuit of Fig. 2-33. Basically, the circuit is an emitter-coupled multivibrator using emitter resistor R2. The feedback path between the output stage and the base of the input stage is by way of resistor R6 and capacitor C3. Long time-constant circuits in both bases permit the formation of a tone output that is reasonably sinusoidal. Check the tone frequency. It will fall somewhere between 1000-2000 Hz. Collector current is approximately 2.5 mA and output about 2 volts maximum. Note the influence of the 50K control on the output level.
2. If you have not permanently attached the components to the vector board, store the board so the components will not be dislodged.

3

Basic IC Circuits

The integrated circuit is an extension of the solid-state science of packing active and passive components into a small package. This has been done with great success. Versatility, reliability, and stability are additional rewards. There is an ever-expanding number of internal circuits and systems, along with external applications, being developed.

Transistors, diodes, and resistors are the only components used in most integrated circuits. Capacitors, while used, are not common because they take up considerable space. Customarily, the internal circuit is designed in such a manner that capacitors are not required.

It is difficult to fabricate a resistor of some precise value into an integrated circuit. Conversely, there is no great problem in including two or more resistors of exactly the same value, even though a certain absolute value is difficult to attain. Hence, internal-circuit systems employ balanced configurations that require equal-value resistors but are noncritical of absolute values. For this reason, the most common integrated circuit is the balanced dc differential amplifier.

BASIC DIFFERENTIAL AMPLIFIER

The differential amplifier is the key circuit of most ICs. The circuit (Fig. 3-1) is basically an emitter-coupled configuration. It has fine stability and a good rejection of undesired signal components. Being direct coupled, no interstage coupling capacitors are needed, providing a saving in space. Correct differential operation requires that the two collector resistances be identical in both ohmic value and general characteristics. In the fabrication of integrated circuits, these conditions are met quite readily and at low cost—in comparison to a similar amplifier designed to use discrete circuit components.

In basic operation, the differential amplifier responds to the signal difference that exists between the base inputs, developing equal-amplitude and out-of-phase collector signals. This type of input is called a differential-mode input signal. In practice, this is accomplished by applying the desired ac signal to one of the base inputs but not to the other. Since no signal is applied to the opposite base, the difference voltage is the signal applied to the input base.

When two equal-amplitude but same-polarity signals are applied to the base inputs, the ac signals across the common-emitter resistor are subtractive. In a situation of perfect balance, the differential amplifier thus performs in a bridgelike manner and there is no output from collector to collector. A much reduced output signal appears from each collector to common.

In-phase signals at the bases are referred to as common-mode input signals. This is usually in the form of any undesired signal, such as hum and interference components. Thus, another advantage of the differential amplifier is its ability to reject and reduce common-mode signals.

In the differential-mode operation, a signal applied to the base of transistor Q1 (Fig. 3-1) appears at the collector output of transistor Q1 and also across the common-emitter resistor. The latter signal component serves as the input signal for transistor Q2. Consequently, a signal of opposite polarity appears at the collector of transistor Q2. In effect, the differential amplifier acts as a phase splitter, developing two equal-amplitude but opposite-polarity signal components at the output.

The differential amplifier has a high order of dc stability, reducing the influence of supply voltage changes, temperature, etc. Supply

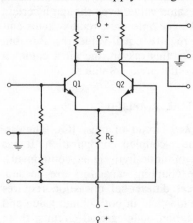

Fig. 3-1. Basic differential amplifier.

voltage variations, in general, create a common-mode type of inter-ference which the differential circuit inherently reduces. Thus, it is a stable dc amplifier and is very practical when designing the multistage circuits that are so common in ICs. Not only does the basic circuit have no need for interstage coupling capacitors, but the emitter bypass is also eliminated.

In summary, integrated circuits have few passive components (re-sistors, capacitors, and coils) and are largely active component sys-tems (transistors and diodes).

STABILITY FACTORS

To repeat, the differential amplifier is a very stable circuit. In un-derstanding its operation, one needs to know about the factors that determine this stability and how special circuit arrangements can further enhance this basic advantage.

The efficiency of the differential amplifier depends on its ability to compensate for any imbalance that may arise. A change in leakage current and/or gain in one side of the differential circuit is balanced out by a like change in the second side. This ability to respond sets the operating limits of the differential amplifier. These limits are best obtained with suitable biasing and proper temperature compensation.

In the rejection of common-mode signals, one depends upon the degenerating effects of the common-emitter resistor. The higher the ohmic value of this resistance, the greater is the rejection. Such in-crease is limited by supply voltage requirements and the difficulty involved in including high-value resistors in integrated circuits.

One answer to obtaining high stability and good rejection of common-mode signals is to include a constant-current emitter source composed of an additional active component rather than a high-value resistor. The fundamental arrangement is shown in Fig. 3-2. In this circuit, the combination of the transistor and its low-value emitter re-sistor acts as the high-resistance constant-current source. The presence of a common-mode signal on the differential transistors affects both voltages and junction resistances. A constant-current source holds the emitter and collector currents constant. In fact, the undesired voltage change appears totally across the constant-current source circuit, which is unbypassed and, therefore, highly degenerative. Thus, the differ-ential gain of the amplifier, in terms of the common mode, is greatly reduced.

Additionally, diodes in the base circuit of this constant-current source provide temperature compensation. When the characteristics of the base-emitter junction of the constant-current transistor and the diode junction are identical, there is exact compensation. An increase in the conductance of the base-emitter junction results from a rise in temperature. Since the compensating diode junction is physically near the transistor, there follows a similar change in its conductance. As a result, a compensation is made in the base bias, keeping the collector-emitter current constant. The basic circuit of Fig. 3-2 is very common in integrated circuits.

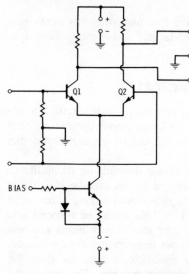

Fig. 3-2. Constant-current source using an active component.

DARLINGTON CONFIGURATION

The previous differential amplifiers have low input impedances. When a high impedance is required, Darlington circuitry is included in the integrated circuits. An additional transistor is included for each pair as shown in Figs. 3-3 and 3-4.

In the normal operation of a transistor, the emitter junction is forward biased and conducts. Input resistance is low and approximates the product of beta times the emitter resistance.

$$R_{IN} = \beta R_E$$

To a degree, the input resistance can be increased by increasing the ohmic value of the emitter resistance. There is a resultant sacrifice in gain and a need for a higher supply potential. The Darlington-pair approach is to use the input resistance of a second transistor as the emitter resistance of the first (Fig. 3-3). The input stage then operates with a highly degenerative emitter circuit and a resultant high-input impedance. Both stages contribute an output with a net gain figure that is comparable to that obtained using a single transistor of the same type but operating with a much lower input resistance. Two such identical circuits are needed for the separate inputs of the configuration shown in Fig. 3-4.

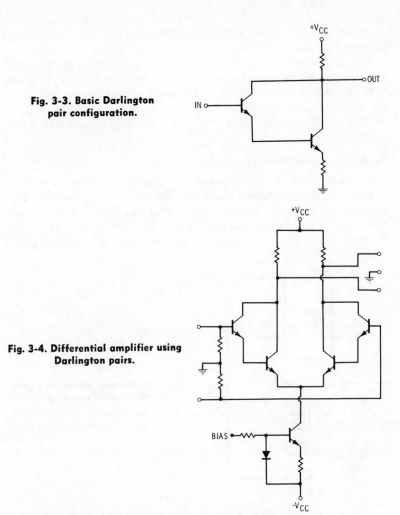

Fig. 3-3. Basic Darlington
pair configuration.

Fig. 3-4. Differential amplifier using
Darlington pairs.

THE DIFFERENTIAL AMPLIFIER IN INTEGRATED CIRCUITS

The advantages of a differential amplifier match the needs of integrated circuits. The IC fabrication process permits the attainment of exceptional balance because all the components of the circuit are processed simultaneously. They have identical characteristics and can be placed near each other. They display similar temperature coefficients and maintain stable electrical characteristics over a wide temperature range. Good input and output isolation are inherent and no neutralization is required. This is basic to simple feedback systems.

Either no capacitors, or very few, are needed in differential circuitry. Large-value resistors can be avoided, and exact, absolute re-

71

sistor values are not critical because differential amplifier performance depends mainly on resistance ratios.

The differential amplifier is a versatile configuration, lending itself well for use as an amplifier, oscillator, limiter, modulator, demodulator, mixer, or several other applications. Its ability to emphasize the desired signal and de-emphasize common-mode components was mentioned previously, along with its capability for highly linear amplification.

Differential Amplifier Current

An example of a simple monolithic differential amplifier is given in Fig. 3-5. The differential transistors and their associated monolithic resistors are identical. This symmetry produces a truly balanced amplifier and is matched so well that there is little circuit unbalance when the two emitters are joined and operated from a common power source.

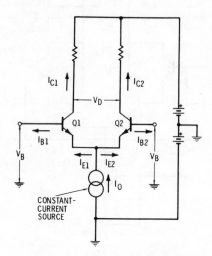

Fig. 3-5. Currents present in a basic differential amplifier.

Zero voltage, or no voltage, base-to-common on each half-circuit produces identical collector currents I_{C1} and I_{C2}. When the amplifier is balanced perfectly and the base voltages are 0, the differential output voltage V_D is also zero. It is zero for any value of equal voltage present at the base of transistor Q1 and the base of transistor Q2.

The sum of the two emitter currents is equal to the source current I_O.

$$I_O = I_{E1} + I_{E2}$$

If current I_O is a true constant-current source, an increase in either emitter current is followed by a like decrease in the other current; the sum remains the same.

When the base voltages of the differential amplifier are unlike, one transistor draws more current than the other. However, the net current I_O remains unchanged. For example, a positive voltage on the base of transistor Q1 causes an increase in its collector current. The rise in the emitter current results in a lower biasing of the second transistor and the collector current of transistor Q2 decreases; so does its emitter current, restoring current I_O to the previous value.

Differential Amplifier Voltage

Inasmuch as the collector current of transistor Q1 increases and the collector current of transistor Q2 decreases, the collector voltage of transistor Q2 becomes higher than the collector voltage of transistor Q1. No longer is there a zero voltage between collector and collector. Instead, the output voltage is positive as measured between the collector of transistor Q2 and the collector of transistor Q1.

If the base voltage of transistor Q1 is made more negative, there is a converse cycle of events. The constant current remains the same, the collector current of transistor Q1 decreases, the collector current of transistor Q2 increases, the collector voltage of transistor Q1 increases, the collector voltage of transistor Q2 decreases, and the output voltage, as measured between collector 2 and collector 1, is negative.

A differential input voltage results in a differential output voltage. This occurs whether the input is a dc voltage change, an ac voltage change, or a combination of both. In terms of the collector-to-collector output voltage, these changes occur on each side of a zero voltage—which is the condition met when the two bias voltages are alike.

Operational Modes

Differential amplifiers can be operated in various modes. The mode described above is referred to as the *differential input, differential output mode.*

The differential amplifier has three additional operational modes (Fig. 3-6). In the circuit shown in Fig. 3-6A, the signal is applied to the base of transistor Q1 and removed at the collector of transistor Q1. A positive input swing results in a negative swing between collector of transistor Q1 and common. This is known as the *single-ended input, single-ended output inverting mode.*

Fig. 3-6B shows the input at the base of transistor Q1 and the output between the collector of transistor Q2 and common. The input swings positive and so does the collector voltage of transistor Q2. Output and input voltages are in phase. This operation is known as the *single-ended input, single-ended output noninverting mode.*

In the final manner of operation (Fig. 3-6C), input voltage is applied from base-to-base and the output voltage can be removed from

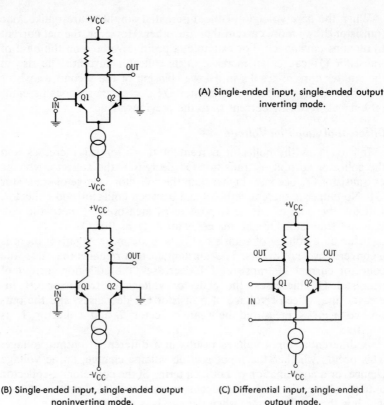

(A) Single-ended input, single-ended output inverting mode.

(B) Single-ended input, single-ended output noninverting mode.

(C) Differential input, single-ended output mode.

Fig. 3-6. Operating modes of a differential amplifier.

either of the collectors. This is known as *differential input, single-ended output mode*.

The various modes of operation require that gain be specified in one of three ways. First, differential-voltage gain is the ratio of collector-voltage change to the difference voltage between the two bases. Second, the double-ended, differential-voltage gain is the ratio of the collector-to-collector voltage change to the base voltage change. Finally, the single-ended differential-voltage gain refers to the ratio of collector-to-common change to the base-voltage change.

Transconductance

The transfer curves of a basic differential amplifier are shown in Fig. 3-7. The center of the linear region corresponds to 0 input voltage. In this region, the transconductance g_m is at its highest. Transconductance, as for other amplifier devices, is the ratio of a small change in output current over the small change in input voltage that produces it. The basic equation for the mutual conductance of the differential amplifier is:

$$g_m = \frac{\alpha I_O}{4KT/Q}$$

where,

α is the current gain,
I_O is the constant current,
K is Boltzman's constant or 1.38×10^{-23},
T is the temperature in degrees Kelvin,
Q is the electron charge or 1.6×10^{-19} coulomb.

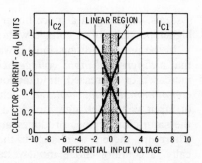

Fig. 3-7. Transfer curves of a typical differential amplifier.

Note from the above that the mutual conductance varies directly with the constant current. This indicates that it is possible to change the mutual conductance and, therefore, the gain of the differential amplifier by varying the value of the constant-current source. Also, the equation indicates that the transfer characteristics and the slope of these curves are a function of alpha and temperature, which are two physical constants. Since these factors are predictable in the case of a monolithic transistor, the mutual conductance equation can be simplified to:

$$g_m \cong \frac{\alpha I_O}{100}$$

Inasmuch as the constant-current (I_O) controls mutual conductance and other operational factors, the differential amplifier is an ideal circuit for use as a mixer, frequency multiplier, modulator, or product detector. When the input is overdriven, it performs well as a limiter, producing a good square-wave output.

The range of linearity and the degree of linearity are improved with some emitter degeneration (Fig. 3-8). Two like-value resistors are placed between the emitters and the constant-current source. This feedback improves the slope of the linear portion of the transfer characteristic and extends it over a greater range of differential input voltages (Fig. 3-9). This technique is widely used in integrated-circuit

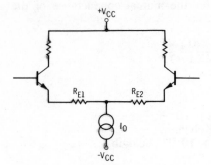

Fig. 3-8. Differential amplifier with emitter degeneration.

systems, resulting in a more linear mutual-conductance change and a higher possible input voltage without distortion.

In the previous discussions, it was assumed, as in normal operation, that a *differential input voltage* existed. When both base voltages are increased or decreased simultaneously, the input signal is known as a *common-mode input voltage*. Such a like-polarity and like-magnitude of input voltage change, provided the transistors are not driven into the saturation region, produces an equal change in emitter currents. Therefore, there is no change in the collector output voltage. Theoretically, a common-mode input-voltage change produces no change in the output voltage of the differential amplifier. How well a differential amplifier rejects these common-mode signal changes is related to the impedance of the constant-current source. Since this impedance is finite, there is always some slight change that results from a common-mode signal. The *common-mode voltage gain* is the ratio of a small output-voltage change that results from a common-mode input-voltage change. In a good differential amplifier circuit, this value is very much less than unity. The ratio of the common-mode gain to the differential gain is known as the *common-mode rejection ratio*.

In monolithic IC forms, the differential amplifier is a well-balanced circuit and, therefore, serves as an excellent dc amplifier, superior to

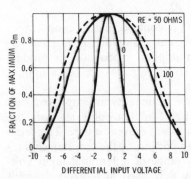

Fig. 3-9. Influence of emitter degeneration on G_m, as a function of input voltage.

other circuit forms. In the perfect form, there is zero voltage between the two collectors when the two base voltages are identical. Perfection is an impossibility and some imbalances do exist, such as unequal-value betas and unequal-value resistors. *Output offset voltage* (V_{OO}) is the term given to the minute difference in dc voltage that exists between collectors when the dc base voltages are identical.

There are two related quantities. The *input offset voltage* refers to the difference in base voltages needed to equalize the two collector voltages. Also, the *input bias offset current* is the difference in the input base-bias dc currents when the dc collector voltages are equal.

CONSTANT-CURRENT SOURCES

The key to top performance from a differential amplifier is its constant-current source. The more one improves this special segment of a differential circuit, the better the overall performance of the amplifier. Various methods for improving the constant-current source will be discussed.

A constant-current source is a high-impedance source. Its effective series resistance is so high that it exerts dominating control over the currents in the two transistors of a differential amplifier and in their collector circuits. The higher the effective impedance of the common source, the less influence the collector current of transistor Q1 exerts in the emitter resistance (Fig. 3-10). Therefore, this current has little

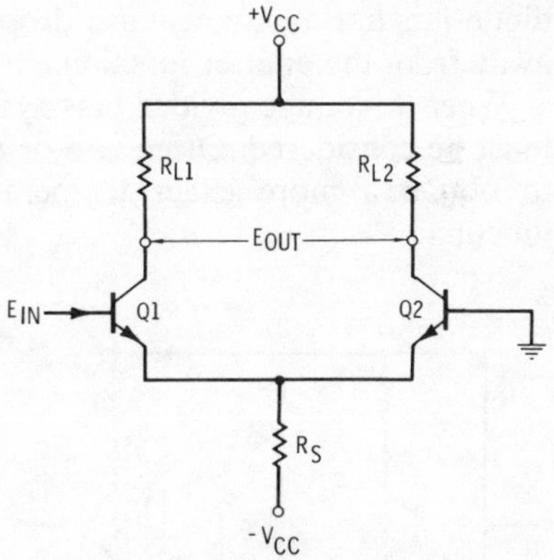

Fig. 3-10. Using a high-value resistor as a constant-current source.

effect on the operation of transistor Q2. The higher the effective emitter resistance R_S, the less influence the collector output circuits have on the input circuits and the operating conditions of the two transistors.

One limit to the ohmic value of the common-source resistance is its influence on the required supply voltage (V_{CC}). The larger the value of R_S, the greater is the supply voltage that is needed to obtain the required value of constant current (I_O). This factor compromises the value of R_S downward.

An alternative is using a transistor as the common-current source, as illustrated in Fig. 3-11. The impedance at the collector of the constant-source transistor is high. Nevertheless, a high-level current (I_O) can be caused in resistor R_E by proper biasing of the base. Resistor R_E can be of low ohmic value and the voltage between the collector of transistor Q3 and common can be correspondingly low. In fact, the voltage drop at the common emitters need only be low enough to equal the sum of the necessary collector-emitter voltage (less than 1 volt) and the voltage drop ($I_O R_E$) across the emitter resistor.

Some inherent temperature compensation exists when the negative-temperature coefficient of the emitter junction (of the source transistor Q3) is matched by the positive-temperature coefficient of the monolithic diffused emitter resistor. In an integrated circuit where the collector resistors are *external,* this compensation is especially helpful because the external collector resistors do not follow exactly the temperature changes that take place within the monolithic structure.

Another temperature-compensating scheme is shown in Fig. 3-12. It is more appropriate when the collector load resistors are diffused types within the monolithic structure. The internal diodes have a compensating response to any change in the emitter-junction bias of the source transistor Q3 caused by temperature and producing a correcting bias change that holds the current I_O constant. A temperature rise at the emitter junction tries to increase junction current. However, the diode junction resistance also drops and some bias current is shuttled away from the emitter junction.

When a voltage-divider bias system is employed, the R1/R2 ratio must be considered. Often two or more diodes are employed in series to obtain a more exact temperature compensation at the emitter junction.

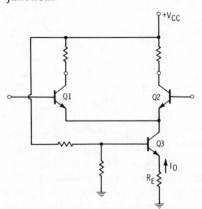

Fig. 3-11   Using a transistor as a constant-current source.

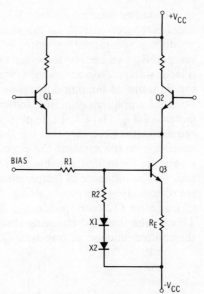

Fig. 3-12. Transistor constant-current source using compensating diodes.

A single diode can be employed as in Fig. 3-13. In this case, the diode and transistor Q3 are matched monolithic transistors within the IC structure. As discussed in Chapter 2, it is common to use a monolithic transistor structure connected as a diode. A special advantage in this circuit is that the two monolithic structures respond in exactly the same manner to temperature change and more exact compensation is feasible.

It can be seen in Fig. 3-14A that the base current finds a path through resistor R3, the emitter junction and resistor R1. Inasmuch

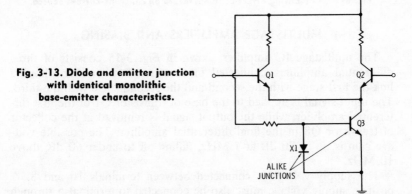

Fig. 3-13. Diode and emitter junction with identical monolithic base-emitter characteristics.

79

as the emitter junction voltage V_{BE} has the dominant influence on the constant-source current I_O, this voltage can vary with temperature and produce a substantial variation in current I_O. Although a high-value resistor R_E can reduce the effects of temperature change, it requires a large voltage drop across the emitter resistor and, thus, is self-defeating, in terms of holding down the supply voltage of its associated differential amplifier. A high-value emitter resistor is avoided with the circuit of Fig. 3-14B. In some arrangements, no emitter resistor is required. It is necessary that the characteristics of the diode-connected transistor be the same as the emitter junction of the constant-current transistor. Note that the bases are biased from the same base-bias source. The influence of temperature produces the same change at both transistors. However, in so doing, transistor Q1 is correcting the bias of transistor Q2 correspondingly, and the current I_O is held constant. This results from the changing shunt influence of transistor Q1 across the emitter junction of transistor Q2.

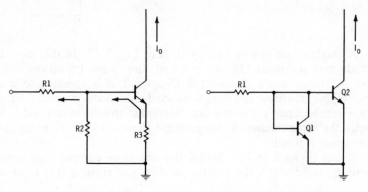

(A) Using a large-value resistor.

(B) Using a transistor connected as a compensating diode.

Fig. 3-14. Combating effect of temperature on constant-current source.

MULTISTAGE AMPLIFIERS AND BIASING

The multistage IC amplifier shown in Fig. 3-15 consists of three differential amplifiers connected in cascade. Two emitter-followers link the first stage with the second and the second stage with the third. The input signal is applied to the base of transistor Q1 of the first differential amplifier, while the output signal is removed at the collector of transistor Q8 in the final differential amplifier. The possible voltage gain is up to 70 dB at 1 MHz, falling off to under 60 dB above 10 MHz.

The supply voltage is connected between terminals 10 and 8. A positive supply voltage must also be connected to terminal 5 through

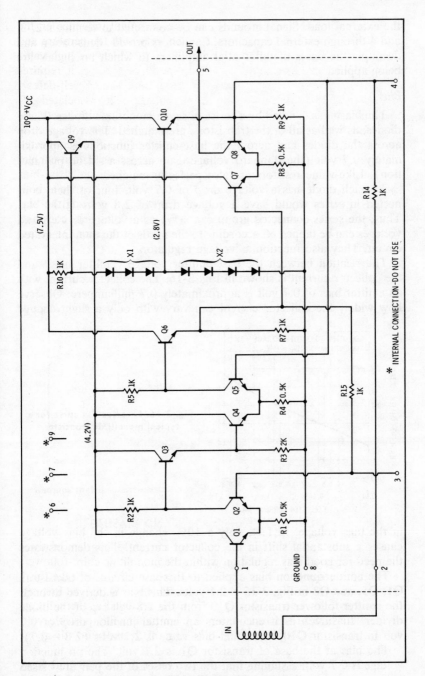

Fig. 3-15. Multistage IC amplifier.

81

the external load. Signal grounds can be connected to terminals 2, 3, and 4 through external capacitors. Capacitors provide dc isolation and decouple the bases of differential amplifiers to which no signals are being applied.

Bias

The biasing is handled by a series of seven monolithic diodes. These diodes, as well as all of the transistors, are matched. The voltage drop across the diodes and across the base-emitter junctions is approximately 0.7 volt (the standard voltage drop across a silicon pn junction). Like-value resistors are also perfectly matched.

As each diode has a voltage drop of 0.7 volt, four of them connected in series would have a voltage drop of 2.8 volts (0.7×4). Thus, the series-connected group can serve as a voltage divider, and voltages can be tapped off according to the needs of the total integrated circuit. They also function as voltage regulators.

The relation between the base-emitter voltage (emitter bias) and the collector current is shown in Fig. 3-16. The collector current with an emitter bias of 0.7 volt is approximately 0.7 milliampere. Observe how widely the collector current can vary with only a slight change

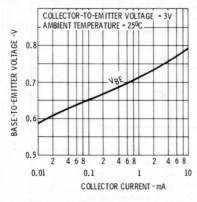

Fig. 3-16. I_c versus V_{BE} curve for a typical monolithic transistor.

in the bias voltage. In fact, only a 10% change in the bias voltage causes a substantial shift in the collector current. This demonstrates the need for good bias regulation within the monolithic chip.

The emitter-junction bias applied to the base circuits of transistors Q1, Q5, and Q8 in Fig. 3-15 is 2.1 volts. This bias is derived through the emitter-follower transistor Q10 from the 2.8-volt tap of the diode divider. Inasmuch as it encounters an emitter-junction drop of 0.7 volt in transistor Q10, the actual bias value is 2.1 volts ($2.8 - 0.7$).

The bias at the base of transistor Q1 is 2.1 volt. The pn junction voltage is 0.7 volt, assuming that the two bases of the pair are biased alike and operate in matched manner. The voltage drop across the resistor R1 will be base bias less the pn junction value or 1.4 volts (2.1 − 0.7). By Ohm's law, the emitter current for the pair is 2.8 mA (1.4V/500Ω). Since the current divides equally between the two transistors, the voltage drop across resistor R2 is 1.4 volts (1000Ω × 1.4 mA). The voltage at the supply-line end of resistor R2 is 4.2 volts (down from the supply-line value because of the presence of emitter-follower transistor Q9).

The above 1.4 voltage drop supplies the necessary base bias for emitter-follower Q3, which direct couples the output of one differential amplifier to the input of the next. Any applied input signal varies about the direct-coupled bias component that is transferred from stage to stage. Likewise, any shift in the dc base bias will be transferred as a dc voltage change to the next amplifier.

Level-Shifting Circuits

One of the problems of a simple direct-coupled amplifier is that progressively higher voltages must be used along the amplifier chain. As a result, the overall supply voltage must be higher than that needed for a capacitive- or transformer-coupled ac amplifier change. Another disadvantage is the fact that at the dc output of the final stage under resting, no-signal condition must be significantly higher than the very low dc level at the input.

The above unfavorable features are overcome with the use of so-called level-shifting circuits. In such an arrangement, it is possible for the dc resting voltage at the input of each stage to be made identical. Furthermore, resting dc voltage at the very output can be made to correspond in value to the dc level at the amplifier input. In the circuit of Fig. 3-15, the emitter-follower transistors Q3 and Q6 are emitter-follower dc level-shifting stages. A simplified version is given in Fig. 3-17. The primary function of a level-shifter is to shift down the average dc level from the output of one stage to the input of the next. Keep in mind that this is the no-signal level (no ac or dc signal input is present).

The technique in the circuit of Fig. 3-17 is to make certain that the ohmic value of resistor R_C, along with the differential constant-current I_O, is such that the dc voltage at the collector output is four forward-biased pn-junction voltage drops above the dc voltage at the input. Under the above condition, the voltage drop at the emitter junction of the level-shifting transistor reduces this figure by one forward-bias junction voltage ($4V_{BE} - 1V_{BE}$). In so doing, the bias at the base (input) of the second differential stage is 2.1 volts, the same as the input-stage bias of the first cascade stage. This arrangement is used between the first and second stages and also between the second and third stages of the integrated-circuit amplifier shown in Fig. 3-15.

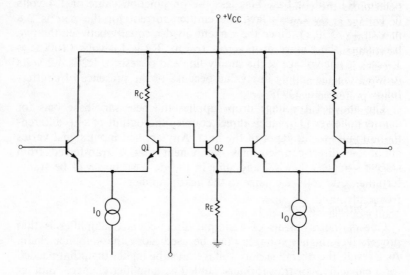

Fig. 3-17. Level-shifting circuit configuration.

A level-shifting arrangement commonly used in the output circuits of monolithic power amplifiers is given in Fig. 3-18. In this arrangement, transistor Q3 serves as the constant-current bias source for transistor Q1. Transistor Q1 is both the driver amplifier and the emitter-follower that supplies signals to the base of the output transistor. The level-shift is accomplished by the voltage drop across resistor R1, as set by the collector current from transistor Q3. In turn, the emitter of the output stage (Q2) has a dc feedback link to the emitter of the control transistor Q3. This feedback path (bootstrap)

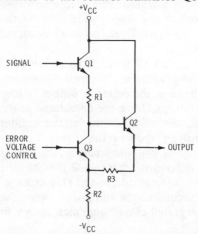

Fig. 3-18. Level shifting in the output stage.

determines the collector current present in resistor R1. The resting dc voltage at the emitter of transistor Q2 can be made the same as the dc voltage at the base of transistor Q1. If we assume that the resting dc voltage at the base of transistor Q1, through previous level-shifting circuits, is the same as at the input of the amplifier chain, the no-signal dc input voltage and dc output voltages can be made zero volts or some other preferred value depending upon design requirements. Further stabilization is accomplished with application of a control error voltage to the base of transistor Q3. Error voltage responds to any undesired common-mode component.

Output Circuits

Integrated circuits employ a variety of output systems. The most frequently used are the so-called uncommitted collector outputs. A good example of an uncommitted collector is the collector of transistor Q8, connected to terminal 5 in Fig. 3-15, which must be connected directly to an external load. This load must present a dc path which

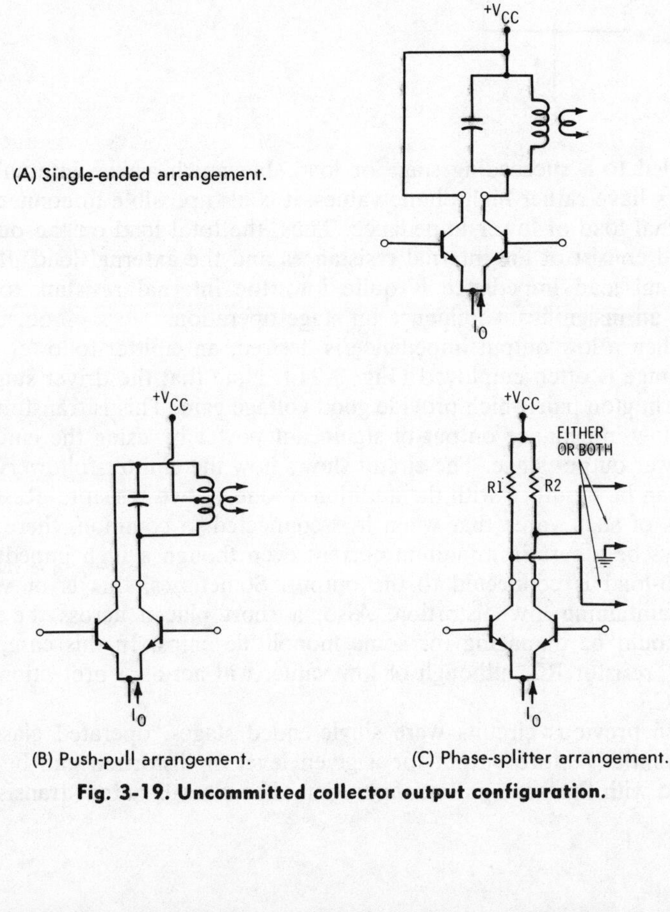

+VCC
(A) Single-ended arrangement.
I0
+VCC
I0
+VCC
EITHER
OR BOTH
R1
R2
I0
(B) Push-pull arrangement.
(C) Phase-splitter arrangement.

Fig. 3-19. Uncommitted collector output configuration.

will supply a proper operating potential for the collector of the output differential amplifier. Quite often, the load is tuned and resonated to the band of frequencies that are to be amplified. In many integrated circuits, both output collectors are uncommitted as shown in Fig. 3-19. Such an output can be connected single-ended, in push-pull, or as a phase-splitter.

Another arrangement is to use internal load resistors as shown in Fig. 3-20. The output can be derived directly and then capacitively

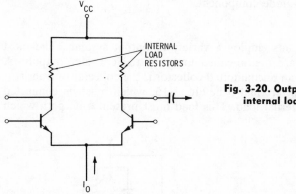

Fig. 3-20. Output circuit using internal load resistors.

coupled to a succeeding stage or load. Inasmuch as the internal resistors have rather high ohmic values, it is also possible to connect an external load of lower impedance. Thus, the total load on the output would consist of the internal resistances and the external load. If the external load impedance is quite low, the internal resistances will have an insignificant influence on stage operation.

When a low output impedance is desired, an emitter-follower output stage is often employed (Fig. 3-21). Note that the driver stage is a Darlington pair which provide good voltage gain. This is transformed to a low-impedance output of significant power by using the emitter-follower output stage. The circuit shows how the emitter-follower output can be modified with the addition of one or two resistors. Resistor R1 is of such value that when it is connected to common, there will always be a certain minimum current even though a high impedance or no-load is connected to the output. Sometimes, this is of value in maintaining low distortion. Also, a short placed across the output could be damaging for some monolithic chips. In this case, the series resistor R2, although of low value, will act as a protection for the IC.

The previous circuits were single-ended stages, operated class-A. Substantially more output for a given level of distortion can be obtained with the push-pull configuration. The class-B output transistors

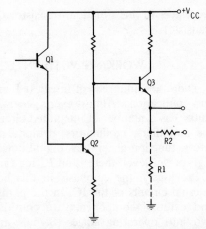

Fig. 3-21. Output stage using an
emitter follower.

of Fig. 3-22 are driven from the same single-ended driver. Note that the input signal to the driver (Q1) is also applied to the base of transistor Q3. After inversion by transistor Q1, the signal is applied to the base of transistor Q2. Since the stages are direct coupled and the no-signal dc output is zero volts, there is very little current in the load.

When the input voltage swings positive from its zero quiescent value, the collector currents of both transistors, Q1 and Q3, increase. In this situation, transistor Q2 is cut off by the drop in Q1 collector voltage. Remember that in class-B operation, activity is switched from one transistor to the other with a change in polarity of the input wave.

As the input voltage swings negatively, collector currents of transistor Q1 and transistor Q3 both decrease. The positive swing in the collector voltage of transistor Q1 causes transistor Q2 to go into conduction and deliver the output for the negative alternation of the in-

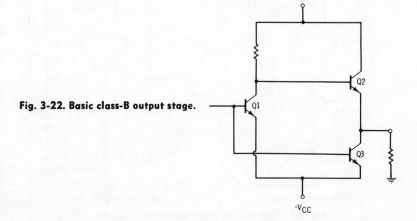

Fig. 3-22. Basic class-B output stage.

put wave. In this situation, transistor Q3 becomes the inactive output transistor.

WORKING WITH INTEGRATED CIRCUITS

There are hundreds of integrated circuits of varying characteristics and applications. Almost every general classification of electronic circuitry has a number of integrated circuits that match its needs. There are amplifiers, modulators, demodulators, phase-comparators, regulators, function generators, digital devices, and so forth. Except at the high-power level, there is an IC for each application.

Previously, the emphasis in this book was on understanding the internal circuits of the IC. In the paragraphs that follow, the internal and external components are considered together by blending the two into operating stages. Signals must be applied and removed; power circuits are needed; and depending upon function, other external components are required.

Voltage

Voltages are derived from either single-supply or dual-supply sources. The basic differential amplifier and constant-current source are shown wired for both types of supplies in Fig. 3-23. In the first circuit (Fig. 2-23A), there are two batteries with common (ground) located between them. The bases operate at common bias level. Shown is resistor-capacitor input, with the ac signal applied to one of the transistor bases of the differential amplifier. Although no signal is applied,

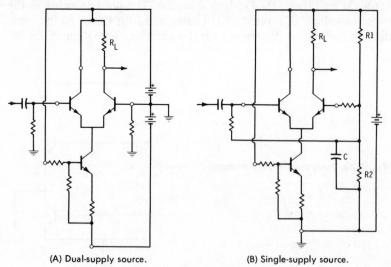

(A) Dual-supply source.   (B) Single-supply source.

Fig. 3-23. Differential amplifier voltage supplies.

the base of the second transistor of the differential pair is connected to common through a like-value resistor. An external load resistor (R_L) is also shown. Signal is removed at this point.

The circuit of Fig. 3-23B shows how the same IC differential amplifier can be operated from a single supply battery. In this case, a resistive voltage divider (R1 and R2), along with a filter capacitor (C), are required. This divider sets the base bias for the differential transistors. The emitter side of the constant-current source is operated at ground potential.

A Typical IC Device

A typical integrated circuit (RCA CA3000) is shown in Fig. 3-24. The terminals are shown by circled numbers. In this case, the inte-

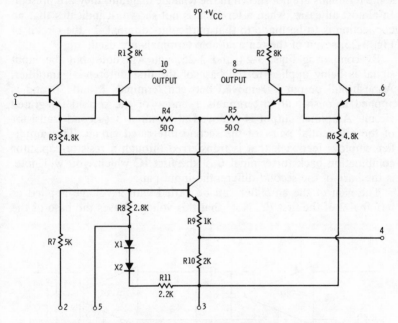

Fig. 3-24. Schematic diagram of the RCA CA3000 integratd circuit.

grated circuit has been packed into a 10-lead TO-5 package. It is basically a differential amplifier using a Darlington input configuration. It employs a transistor constant-current source along with two temperature-compensating diodes. Application is mainly as a linear amplifier. The device is capable of handling a good signal-input level because of the use of individual emitter resistors in the differential amplifier. Note that internal resistor values are indicated on the schematic.

89

Two of these ICs, connected in a cascaded, resistance-capacitance-coupled amplifier (with feedback), are shown in Fig. 3-25. Note that a triangle represents the integrated circuits and the internal components are not shown. Integrated circuits are shown in this manner in most schematics.

In most applications, it is not necessary to show the internal wiring because it cannot be changed in any way. However, in learning, designing, or simple experimentation, it is often helpful to know just what is inside the chip, so you can make circuit changes (or explore other possibilities) of value for your particular application. In this case, the diagram supplied by the manufacturer of the device should be consulted.

Terminal numbers of the IC are indicated in Fig. 3-25. Although some terminals are not shown in the triangle diagram, they are present. In almost all cases, when a terminal is not shown, it indicates that no connection is to be made to that particular terminal. In the circuit of Fig. 3-25, seven of the ten available terminals are used.

By comparing Figs. 3-24 and 3-25, you will note that the input signal is being applied to the base of the first differential amplifier. A push-pull output is removed between terminals 8 and 10 and is applied in push-pull to terminals 1 and 6 of the second integrated circuit. Amplifier output is removed at terminal 8 (second transistor of the differential pair) of the second integrated circuit. This amplifier employs feedback that is transferred through a resistor-capacitor combination back to terminal 6 of the first IC which, you will note, is the base of the second differential-input pair.

The gain of the amplifier can be varied with a voltage applied to terminal 2 of the first IC. Note that this voltage biases the base of the

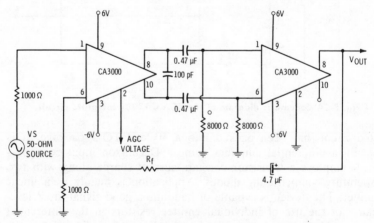

Fig. 3-25. Resistance-capacitance-coupled amplifier using two integrated circuits.

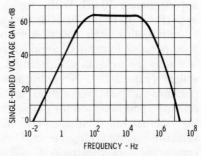

Fig. 3-26. Graph showing amplifier
frequency response.

constant-current transistor. As discussed previously, the base voltage
controls the constant current (I_O) and, thus, can regulate the gain of
the differential amplifier. If the gain is to be controlled automatically,
this voltage can be derived from an automatic-gain-control system. The
gain and frequency response of the cascaded amplifier is shown by the
graph of Fig. 3-26. Over the flat region, the gain is in excess of 60 dB.
Note that the response extends upward into the megahertz regions
even though no peaking coils or resonant circuits are employed.

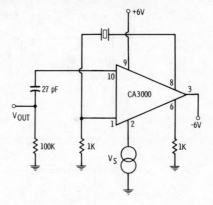

Fig. 3-27. Crystal-controlled
oscillator circuit.

Applications

A single IC of this type can operate as an efficient oscillator. A
crystal-controlled circuit is shown in Fig. 3-27. The feedback path
exists between the collector of the second transistor in the differential
pair to base one of the first. Output is removed from the collector circuit of the first transistor in the differential pair. If desired, the oscillator frequency can be modulated with a low-frequency tone applied
to the base of the constant-current transistor. This in turn modulates
current I_O.

The CA3000 integrated circuit can be operated as a radio-frequency
amplifier as shown in Fig. 3-28. This 10-MHz amplifier has a gain of

approximately 30 dB. With appropriate tuned circuits, it can be made to perform up to the 30-MHz range. Input is single-ended to terminal 1; output, also single-ended, is removed at terminal 8.

A popular integrated circuit, represented by the RCA CA3028A, is simply a single-stage differential amplifier and its constant-current source (Fig. 3-29). No unbypassed emitter resistors or diode temperature-compensating diodes are included. There are no collector resistors, and the collector output terminals provide means for attaching external tuned circuits. This IC does not offer the advantages of versatile biasing arrangements. However, the simple arrangement and compactness of construction make it an ideal device for use as a radio-frequency or intermediate-frequency amplifier. Performance is

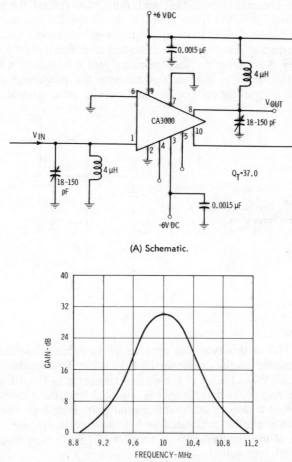

(A) Schematic.

(B) Response curve.

Fig. 3-28. Narrow-band tuned amplifier.

good up into the vhf frequency range; the device can even serve as a front end for an fm receiver.

Wiring and supply voltage configurations for various applications of the CA3028A are given in the circuits of Fig. 3-30. Operation as a differential radio-frequency amplifier is shown in Fig. 3-30A. Single-ended signal is applied to terminals 1 or 5 (bases of the differential amplifier). A single power supply is used with a resistor voltage divider supplying the correct base bias. Output is removed single-ended from one of the differential collectors. Since a differential amplifier circuit has good limiting action, this configuration can also be used as a limiter in fm i-f systems.

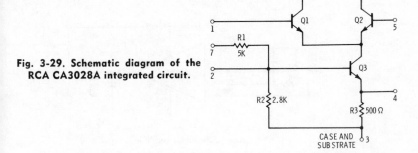

Fig. 3-29. Schematic diagram of the RCA CA3028A integrated circuit.

For high-gain applications, the cascode circuit of Fig. 3-30B is preferred. Signal is applied to terminal 2 of transistor Q3 which acts as the input transistor of the cascode pair. Note in Fig. 3-29 that the collector of transistor Q3 is direct coupled to the emitter of transistor Q2. Output is removed at terminal 6 which is the collector of transistor Q2. Biasing is again handled with a two-resistor divider. If desired, agc voltage can be applied to terminal 7. This connection biases the base of transistor Q3 through monolithic resistor R1.

Fig. 3-30C shows the amplifier operated as an oscillator. The feedback path, by way of capacitor C_f, is between terminals 6 and 1, or from the collector of the output transistor to the base of the input transistor. The level of oscillation can be controlled by regulating the terminal 7 bias voltage, which sets the level of the constant-source current.

To use this oscillator arrangement as a converter, illustrated in Fig. 3-30D, it is only necessary to apply the rf signal to terminal 2 which is the base of the constant-current transistor Q3. In this circuit, the oscillator feedback path is to terminal 1 or the base of transistor Q1. Mixing action takes place in transistor Q1 producing a difference frequency at terminal 8. The oscillator tuned circuit is connected to terminal 6 (the collector of transistor Q2).

A mixer is shown in the circuit of Fig. 3-30E. A balanced rf signal is applied between the two bases, terminals 1 and 5. The oscillator component is applied to the base of the constant-current transistor. Mixing takes place in the balanced differential arrangement, and the difference-frequency signal is removed in the push-pull output connection using terminals 6 and 8 (both collectors of the differential amplifier).

An fm tuner using two RCA CA3028A integrated circuits is shown in Fig. 3-31. The input stage of the tuner functions as an rf amplifier and is connected differentially to obtain the best noise performance. Biasing of the constant-current source occurs at terminal 7 by way of resistor R1. The level is adjusted to obtain a power gain of 15 dB. A similar arrangement is used to bias the two bases of the differential

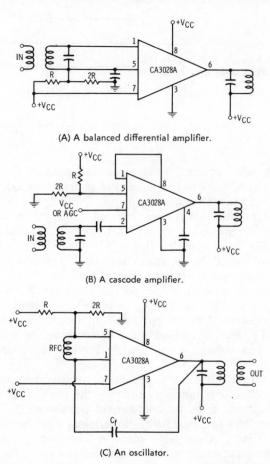

(A) A balanced differential amplifier.

(B) A cascode amplifier.

(C) An oscillator.

Fig. 3-30 Typical circuits

amplifier. Output is derived at terminal 6, the collector of the second differential pair. Note that for proper impedance matching, both the input coil L1 and the output coil L2 are tapped because of the low input and output impedances of the transistor circuit.

A double-tuned transformer links the amplifier with the converter. This transformer also serves as the tuned circuit for the oscillator, with the feedback path by way of the 1-pF capacitor that links the base circuit (terminal 1) to collector terminal 8. The incoming signal is applied to the base of the constant-current transistor at terminal 2. The difference frequency is removed at collector terminal 6 and applied to the intermediate-frequency transformer.

AUDIO POWER AMPLIFIERS

A popular example of a monolithic IC audio-power amplifier is the RCA CA3020 type shown in Fig. 3-32A. Typical specifications are given in Fig. 3-32B. Maximum power output using the CA3020A version approaches 1 watt. This maximum is obtained with an input signal of approximately 45 millivolts. Note that the maximum signal current is in excess of 140 mA. Although used widely as an audio-

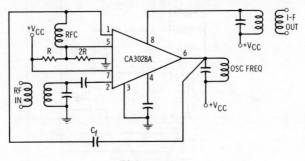

(D) A converter.

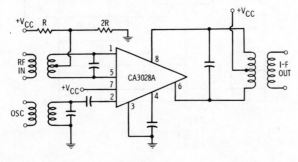

(E) A mixer.

using the RCA CA3028A IC.

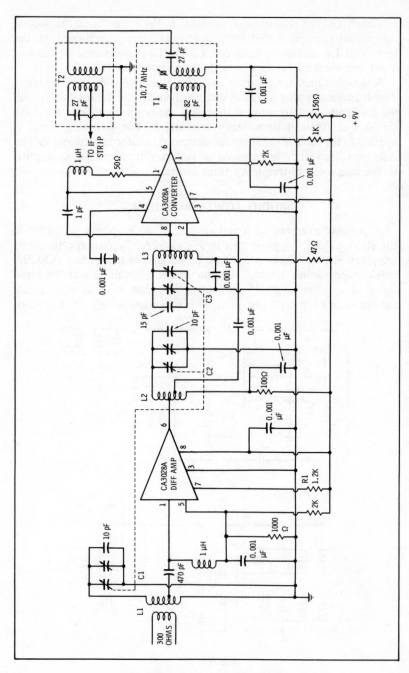

Fig. 3-31. Fm tuner using two ICs.

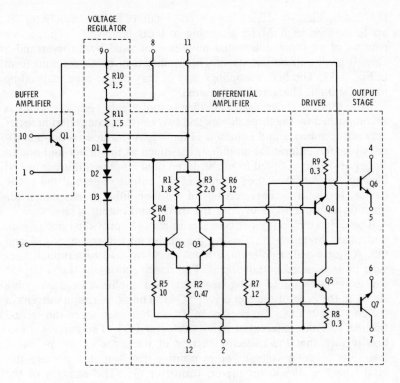

(A) Internal circuitry.

CHARACTERISTIC		CA3020		CA3020A	
POWER SUPPLY— $V-_1$		9		9	V
$V+_2$		9		12	V
ZERO-SIGNAL IDLING CURRENT— I_{CC1}		15		15	mA
I_{CC2}		24		24	mA
MAXIMUM-SIGNAL CURRENT— I_{CC1}		16		16.6	mA
I_{CC2}		125		140	mA
MAXIMUM POWER OUTPUT AT 10% THD		550		1000	mW
SENSITIVITY		35		45	mV
POWER GAIN		75		75	dB
INPUT RESISTANCE		55		55	kΩ
EFFICIENCY		45		55	%
SIGNAL-TO-NOISE RATIO		70		66	dB
% TOTAL HARMONIC DISTORTION AT 150 mW		3.1		3.3	%
TEST SIGNAL		1000 Hz / 600Ω GENERATOR			
EQUIVALENT COLLECTOR-TO-COLLECTOR LOAD		130		200	Ω
IDLING-CURRENT ADJUST RESISTOR (R11)		1000		1000	Ω

(B) Typical specifications.

Fig. 3-32. IC power amplifier (RCA CA3020).

power amplifier, the device has a voltage gain of approximately 60 dB, up to as high as 8 MHz, assuming a 3 dB bandwidth. The device consists of an input differential amplifier followed by a driver and a class-B push-pull output stage. This is shown in block diagram form in Fig. 3-33. The buffer amplifier may or may not be used, depending upon the input characteristics desired.

Note that the differential amplifier is thoroughly regulated. This is accomplished by the three diodes and two resistors connected in series between terminal 9 and common (Fig. 3-32A). These monolithic diodes bias the differential amplifier. Inasmuch as the driver and output stages are direct coupled from the differential amplifier, this controlled biasing also sets the idling current of the output stage and the entire device. In fact, the characteristics of the monolithic diodes match the junction characteristics of both the driver and output stages.

The differential amplifier operates in class-A, providing power gain and phase inversion for driving the push-pull driver stages Q4 and Q5. A single-ended differential input signal applied to terminal 3 results in a balanced output signal. The positive swing at transistor Q2 causes its collector to swing negatively and applies a negative-going signal to the base of the top driver (Q4). The same rise in current in the emitter circuit is also present in the emitter junction of the second transistor of the differential pair (transistor Q3). The current direction is such that the collector current of transistor Q3 drops. As a result, its collector voltage swings positive, resulting in a positive rise at the base of the lower driver, transistor Q5. The emitters of the driver stage supply equal-amplitude but opposite-polarity drive signals to the bases of the output stage. The output resistors Q6 and Q7 are monolithic power transistors and are connected to operate class-B when the emitter terminals 5 and 6 are connected to common (ground). Push-pull output is derived between terminals 4 and 7.

Two typical circuits are shown in Fig. 3-34. The unit is so completely self-contained that few external components are needed. In

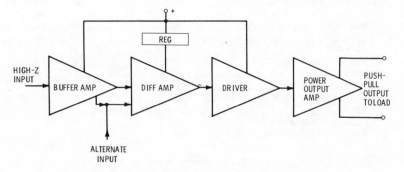

Fig. 3-33. Functional block diagram of schematic shown in Fig. 3-32.

the circuit of Fig. 3-34A, the signal is applied directly to the base of the differential amplifier by way of terminal 3. In this connection, the input resistance is approximately 700 ohms.

A 50,000-ohm input resistance is obtained using the circuit of Fig. 3-34B. The signal is applied to the base of the buffer amplifier by way of terminal 10. Externally, through potentiometer R2, the emitter of the buffer amplifier is connected to the base input of the differential amplifier. This emitter amplifier connection provides the higher input resistance. The buffer amplifier is biased by connecting resistor R1 to the supply voltage. Frequency response of the amplifier is determined by the values of capacitors C1 through C5. Capaci-

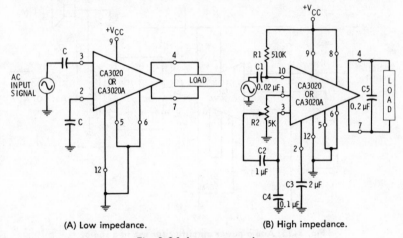

(A) Low impedance.   (B) High impedance.

Fig. 3-34. Input connections.

tors C1, C2, and C3 influence the low-frequency response. Note that these are mainly interstage coupling capacitors and influence low-frequency results just as the interstage capacitors of the ordinary resistance-capacitance-coupled amplifier do. Capacitors C4 and C5 shunt the signal path and, like distributed capacitance in any amplifier, determine the high-frequency response and roll off. The capacitor values indicated establish the restricted voice-communications bandwidth of 300 cycles to 3000 cycles. For other responses, you can increase or decrease the above capacitor values.

The amplifier of Fig. 3-35 provides more than ½-watt output for driving a low-impedance speaker. Collector-to-collector load impedance falls somewhere between 130 and 200 ohms. Thus, the push-pull output transformer must match this amount to the 3.2- or 8-ohm speaker. Normal operating conditions would be an idling current of 22 mA and an input resistance of 50,000 ohms. Maximum input voltage of 45 millivolts will drive the amplifier to maximum output. An even higher output can be obtained by using a 12-volt supply and the CA-3020A device.

A similar circuit connecting the device for use as a simple intercom system is shown in Fig. 3-36. In this arrangement, an input transformer is also needed to match the speakers (being used as a microphone) to the higher input resistance of the device. By inserting the additional resistor between the supply voltage point and terminal 11, there is an additional injection of current into the voltage regulator circuit, and the idling current can be increased. A somewhat higher output can be obtained.

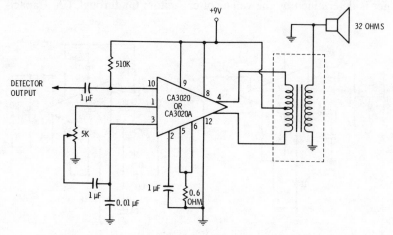

Fig. 3-35. Audio power amplifier for use in a radio.

The SL402/SL403 power amplifier ICs (Plessy Semiconductors, England) differ from most in that they do not employ differential amplifiers (Fig. 3-37). Stability depends upon feedback and stiff voltage-regulating circuits. The SL402 is a two-watt output, audio amplifier; the SL403 is a three-watt amplifier.

The IC is a complete amplifier too, including a preamplifier as well as a main amplifier. The preamplifier can be wired into or left out of the circuit through an external terminal. A jumper connected between terminals 4 and 5 connects the preamplifier into the circuit.

A pair of Darlington trios serve as input for the preamplifier and the main amplifier. For the preamplifier, input transistors Q1 and Q2 are cascaded emitter-followers while transistor Q3 is the common-emitter output. Such an arrangement has a high input impedance (in the megohm range).

The transistor trio forming the main amplifier is followed by two cascaded common-emitter stages, Q7 and Q8. These prevent the output stage from placing a low-impedance load on the output of the

100

Darlington voltage amplifier. The class-B output transistors are Q15 and Q16. Diodes are used to channel the currents between the two output transistors, responding to either the negative or positive alternation of the drive wave applied to the base of transistor Q16.

Considerable negative feedback insures a linear response and a distortion as low as 0.3% for a one-watt output. Shunt feedback exists from the output through a 1K resistor to the emitter of transistor Q6 of the main amplifier input stage.

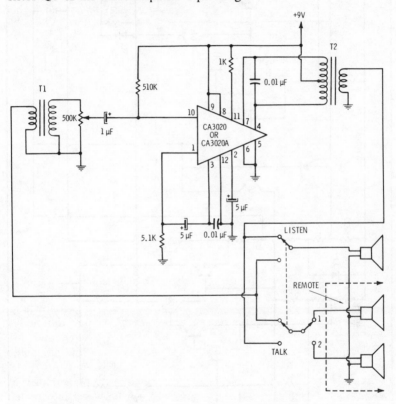

Fig. 3-36. A single-IC intercom system.

A decoupling facility is brought out to terminal 7. A large capacitor, connected between this point and common, permits reducing the hum to a very low level.

Diode and transistor arrangements provide over-voltage protection. Excessive-current protection is provided by the monolithic silicon controlled rectifier X2. If the voltage across either current-monitoring resistor is exceeded, the SCR fires, shuts down transistors Q9 and Q10 which, in turn, switch off the output amplifiers.

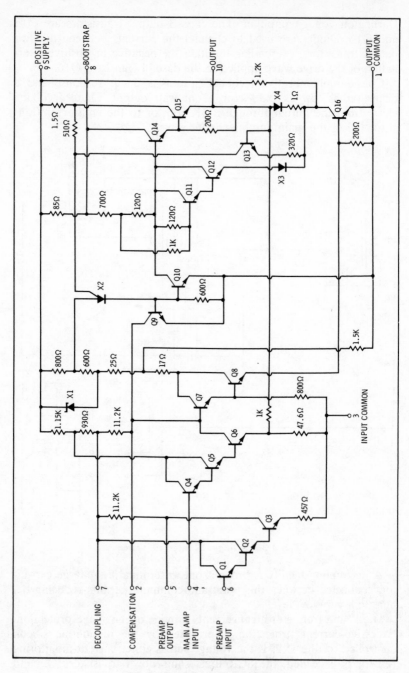

Fig. 3-37. Audio power amplifier.

102

A high-gain voltage amplifier is shown in Fig. 3-38. An input signal of only 25-mV rms results in maximum power output. A 1-megohm potentiometer serves as the audio gain control. A potentiometer connected to terminal 7 provides bias adjustment. This is set for maximum undistorted output when you wish to reduce distortion to the lowest possible value.

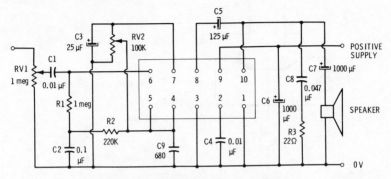

Fig. 3-38. High-gain voltage amplifier.

Note that the output of the preamplifier is connected to the input of the main amplifier by placing a short circuit between terminals 4 and 5. Various high-value decoupling capacitors are used externally to keep the hum at a very low level and to prevent feedback problems.

EXPERIMENT 2: DIFFERENTIAL AMPLIFIER

General

The differential amplifier is the basic stage of most integrated circuits. It amplifies the difference voltage between its two inputs (Fig. 3-39). Customarily, one of the inputs is grounded and a signal is applied to the opposite input as in Fig. 3-39A. An alternative to the single-ended input is the application of opposite-polarity signal components to the two inputs as shown in Fig. 3-39B. These two circuits show the operation of the amplifier in terms of a differential-mode input signal.

When a signal is applied to two inputs connected in parallel as in Fig. 3-39C, there will be either no output or a minimum output. This is called common-mode operation, and a major advantage of the differential amplifier is its ability to reject common-mode components. The output from a differential amplifier can be removed either balanced or single-ended as shown in Fig. 3-40. The balanced arrangement of the circuit in Fig. 3-40A provides equal-amplitude but opposite-polarity outputs. As shown in Fig. 3-40B, single-ended output can be derived from either side.

The HEP 580 integrated circuit can be connected in various differential modes. This will be the objective of our experiment. Both dc and ac checks will be made. The differential amplifier will be built around the second IC socket mounted on the experiment board.

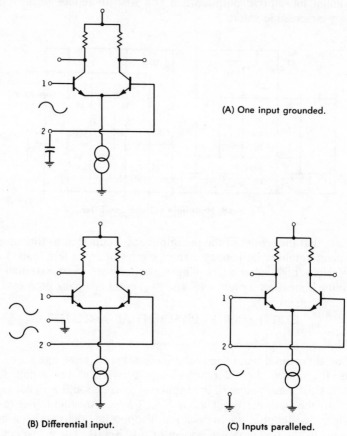

(A) One input grounded.

(B) Differential input.

(C) Inputs paralleled.

Fig. 3-39. Input configurations.

Procedure 1: Construction and DC Checks

1. Construct the circuit of Fig. 3-41. The transistors of each pair are connected in parallel when setting up the differential circuit. Thus, there are two inputs and outputs brought out to binding posts. To make it convenient in checking out the various input and output modes, you can also locate a common (grounded) binding post between the input and output posts as shown. Supply voltage is a 9-volt transistor radio battery. A 1½-volt battery is needed for checking out the dc performance of the differential amplifier.

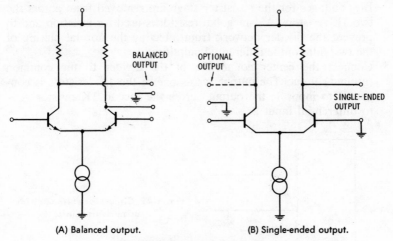

(A) Balanced output.   (B) Single-ended output.

Fig. 3-40. Output configurations.

2. Apply power and measure the total collector current and the individual collector voltages. Normally, the collector current will be a bit under 1.5 mA and the individual collector voltages somewhat greater than 6 volts.
3. Connect the voltmeter from collector to collector. This reading will be about 0 volts. On the 0.4-volt scale of a typical high-impedance voltmeter, the reading may be about 0.15 volt. The imbalance is likely to be due to a slight difference in the ohmic values of the 1-megohm bias resistors.
4. To make the dc voltage checks, assemble the voltage divider network shown in Fig. 3-42. It is connected across a 1½-volt battery.

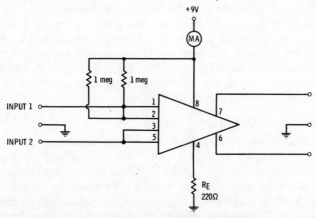

Fig. 3-41. Circuit for differential dc measurements.

Bias voltages for the transistor stage are removed from across the two 1K resistors. A 1-megohm resistor is used for isolation and to prevent the divider network from affecting the normal biasing of the two sides of the differential amplifier.

5. Connect the center connection of the divider to the common (ground) input. The output across one of the 1K resistors is connected to input 1; the output across the second 1K resistor goes to differential input 2.

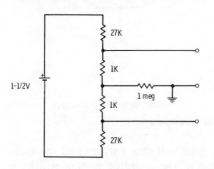

Fig. 3-42. Circuit for voltage divider dc measurements.

6. Turn on the amplifier and measure the voltages for each collector-to-common and then from collector-to-collector. Note that one collector voltage reads less than the value taken in Step 2; the other, higher. The total collector current remains unchanged with the current being less in one section but more in the other, producing the same total current. The voltage measured from collector-to-collector is the difference between the two collector-voltage readings. This is the balanced output.

7. Reverse the battery polarities and repeat Step 6. The collector voltages will be the same but interchanged. The voltage from collector-to-collector will also be the same but of opposite polarity.

8. Connect a jumper between the two inputs. Connect the positive side of the potential (as measured across one of the 1K resistors) between this point and common. There will be a low voltage measured collector-to-collector. This is the common-mode output and is, of course, much lower than the output that is obtained with the differential application of dc voltage.

Procedure 2: AC Operation

1. Disconnect the battery divider from the input and reconnect the circuit as shown in Fig. 3-43A. The single-ended output of an audio generator set for a 1000 Hz frequency is connected to supply a differential signal to the amplifier input of the circuit in Fig. 3-43A. Note that input 2 is grounded through a 0.1-μF capacitor. The high side of the audio generator output is applied to input 1.

106

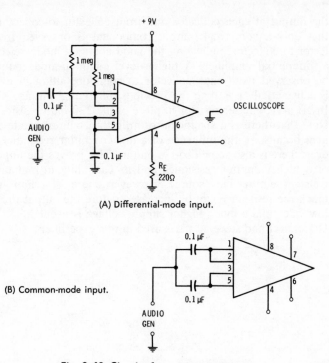

(A) Differential-mode input.

(B) Common-mode input.

Fig. 3-43. Circuits for ac measurements.

The output indicator is an oscilloscope connected to one of the collector outputs.

2. Turn on power. Note that the total collector current is a bit less than 1.5 mA, just as it was for the previous procedure. Slowly increase the output of the audio generator up to the point where the sine wave as displayed on the oscilloscope begins to distort. Back off adjustment slightly to obtain an undistorted output.

3. Transfer the oscilloscope leads to the other collector output. Note that the two outputs are of approximately the same amplitude. Now, connect the oscilloscope to read the output from collector-to-collector. Note that the output is twice that observed at the individual collectors.

4. Transfer the input, grounding input 1 and connecting the high side of the audio generator output to input 2. The results are the same as above.

5. Apply the audio signal generator in common-mode fashion. The connection is shown in Fig. 3-43B. Note that the inputs are paralleled and connected to the high side of the audio generator output. The ground side of the audio generator remains connected to the ground of the differential amplifier. Turn on power and observe

the output at each collector and from collector-to-collector. Note that only a very weak output component is observed from collector-to-collector, indicating the good common-mode rejection of a differential amplifier. A higher-level common-mode output can be observed from each collector to common, although even this is substantially less than the previous differential output.

6. Insert a 1K emitter resistor in place of the 220-ohm resistor (R_E). Note that there is a significant drop in the common-mode component because of the influence of a higher emitter resistance. However, there is also some drop in output. This proves the importance of a higher emitter resistance and its capability in maintaining a constant emitter-bias source. However, there is a limit to how much the resistance can be increased before the output drops to a low value and a much higher supply voltage is needed.

7. Disconnect and store all parts used in this experiment.

4

Operational Amplifiers

An operational amplifier is a special form of linear integrated circuit with high dc and ac gain and a high stability. Its many applications include digital as well as linear functions. Some of these are signal amplification, waveform generation and shaping, analog-digital conversion, impedance transformation, instrumentation, and so forth. It can also perform mathematical operations (the original purpose of this device) that include summations, subtractions, integration, and differentiation.

When the amplifier no-feedback gain (open-loop gain) is adequate and a feedback system is of correct design, the closed-loop gain and characteristics of the operational amplifier become a function of only the feedback components. Basically the relative ohmic values of two external resistors can be used to set the operating characteristics for a particular operational amplifier.

BASIC OPERATION

The operational amplifier or *op amp* is a multistage affair that includes the popular monolithic differential amplifier as well as other types of amplifiers, level shifting, and matching and voltage regulation stages. A typical operational amplifier is shown in Fig. 4-1.

It consists of a pair of differential amplifiers and a cascaded single-ended output stage. The first differential amplifier uses transistors Q1 and Q2 along with the constant-current transistor Q6. There is also a temperature-compensating diode X1. The second amplifier consists of transistors Q3, Q4, Q7, and diode X2.

The feedback circuit of transistor Q5 reduces any common-mode error signal. It does this by evaluating the signal at the emitters of transistors Q3 and Q4. When the second differential amplifier is driven push-pull, there should be a zero level at this point, indicating proper operation of both the first differential amplifier and the input system of the second. If an error voltage is present, a correcting voltage is developed across resistor R2 (in the differential collector circuit) by transistor Q5. This same transistor (Q5) also introduces an error bias into the constant-current transistor circuit (Q7) to bring a further reduction in common-mode signal.

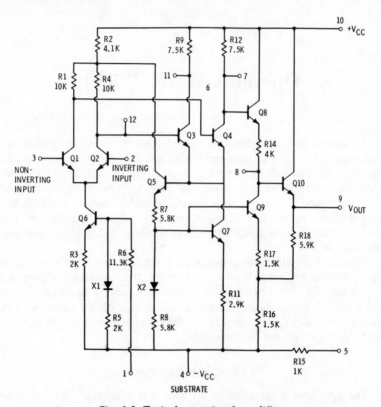

Fig. 4-1. Typical operational amplifier.

The emitter circuit of this important stabilizing transistor (Q5) also supplies a dc error voltage to the base of transistors Q7 and Q9. Common-mode dc stabilization results. For example, a decline in supply voltage reduces the base biases at transistors Q7 and Q9. Since the collector of Q9 is linked to the base of transistor Q10 and to the emitter of transistor Q8, there is a similar change there as well as at the bases of transistors Q3 and Q4. The net result is to produce increases in collector voltages that compensate for the supply voltage decline. The feedback stabilization furnished by transistor Q5 offers excellent common-mode rejection, tolerance to supply voltage change, and high open-loop stability.

The output of the second differential amplifier supplies drive to the base of the emitter-follower transistor, Q8. In turn, transistor Q8 supplies a signal to the base of the single-ended emitter-follower output transistor, Q10. There is a limited amount of signal gain contributed by the output circuit as the result of the *bootstrap* (small amount of positive feedback) from the emitter of transistor Q10 to the emitter of transistor Q9. Therefore, transistor Q9 serves a dual purpose; as a constant-current source for the drive transistor and as a part of the bootstrap. The output system also provides a dc level shift and the level at terminal 9 now corresponds to the no-signal input level.

The RCA CA3015 op amp, connected as a single-supply 10-dB amplifier, is shown in Fig. 4-2. The ac signal is applied to the base of

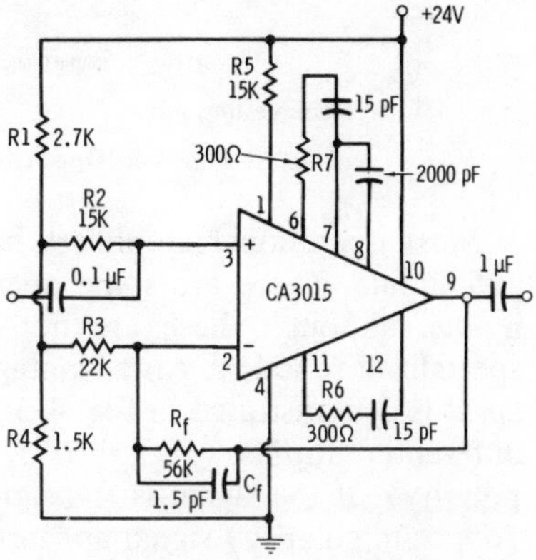

Fig. 4-2. Amplifier application of an operational amplifier.

the first transistor of the first differential pair. However, dc bias is applied to both bases by way of the four-resistor bias network—R1, R2, R3, and R4. Output is taken off at terminal 9, the emitter of output transistor Q10. The feedback component is also derived here and is transferred back to the second base (terminal 2) of the input differential amplifier through the feedback resistor-capacitor combination, R_fC_f.

Supply voltage is applied directly to the collector of the output transistor by way of terminal 10. Supply voltage for the base of the input constant-current transistor Q6 is supplied through resistor R5. High-frequency stabilization is provided by resistor R6 and its associated capacitor connected between collector and base of the differential transistor Q3. Resistor R7 and its associated series capacitor performs in a similar capacity for transistor Q4 of the second differential amplifier. Terminal 4 connects to common.

In summary, the feedback plan of the operational amplifier provides high stability and characteristics that can be, if desired, related directly to the feedback network. An operational amplifier has additional favorable attributes. Some have high ac and dc voltage gain, from several thousand up to as high as a million. The bandwidth of some extends uniformly from dc up to several hundred megahertz. Common-mode rejection is excellent and there is little dc offset or drift with temperature. If desired, the input impedance of some types can be made very high, and the power input and current requirements made insignificant.

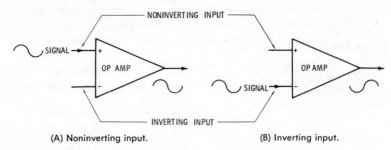

(A) Noninverting input.   (B) Inverting input.

Fig. 4-3. Operational amplifier inputs.

Most operational amplifiers have single-ended outputs and push-pull inputs. There are some operational amplifiers with single-ended inputs, although these are not as versatile and may have only a specialized function. An advantage of the differential input in this respect is demonstrated in Fig. 4-3. When signal is applied to one of the differential inputs (Fig. 4-3A), the output is not inverted (same polarity). If the signal is transferred to the second input, an inverted (opposite polarity) signal appears at the output (Fig. 4-3B). In terms of an ac input signal, this means that an input can be selected to produce either an in-phase or an out-of-phase output signal. Ordinarily, the input terminals of operational amplifiers are labeled as either the inverting input or the noninverting input.

OPERATIONAL PRINCIPLES

Several mathematical relations aid in understanding the characteristics of an operational amplifier. Such an amplifier has open-loop intrinsic input and output impedances and an inherent voltage gain A_O, called the open-loop differential voltage gain. The latter quantity is frequency related as illustrated in Fig. 4-4. As a function of various supply voltages, the curves show the gain-frequency characteristics for the RCA CA3015 monolithic operational amplifier. Note for the ±6 volt operation condition that the open loop gain is 60 dB. (60 dB

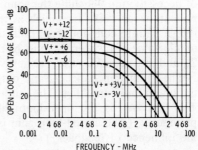

Fig. 4-4. Op amp open-loop voltage-gain curves.

corresponds to a voltage amplification of 1000.) At about 300 kHz, the response is down 3 dB, while near 3 MHz the gain is down 20 dB. Unity gain falls at 18 MHz.

Fig. 4-5 demonstrates the excellent common-mode rejection of an operational amplifier with a differential input. This rejection is in excess of 90 dB for frequencies below 100 kHz. The rejection at 10 MHz remains above 55 dB.

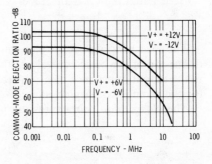

Fig. 4-5. Op amp common-mode rejection-ratio curves.

Phase compensation maintains the high stability of an op amp and removes the tendency for self-oscillation at high frequencies, particularly near the unity-gain region. At these high frequencies, the phase angle approaches the point at which there is a turnover to positive feedback instead of the desired negative feedback. External components connected in a series resistor-capacitor combination provide phase compensation and a more linear decline in the gain-bandwidth factor at high frequencies (Fig. 4-6). The dropoff is maintained reasonably constant at a figure of 6 dB per octave. This phase compensation is the function of the networks connected between terminals 6 and 7, and between terminals 11 and 12 in Fig. 4-2.

Equivalent diagrams of inverting and noninverting operational-amplifier configurations are given in Fig. 4-7. These are closed-loop equivalents with a portion of the output transferred back to one side of the differential input by way of resistors Z_F and Z_R. Resistor Z_R,

113

of necessity, must include the source impedance as well as bias components when used. Inasmuch as there is a differential input, the dc return path to common from both sides must be equal. Therefore, the ohmic value of resistor R_R must be made equal to the ohmic resistance Z_R and must consider the resistance of the signal source V_{IN}.

The input impedance Z_i and output impedance Z_O are intrinsic to the operational amplifier. The output generator has a value of $A_O V_{IN}$. The term A_O is the open-loop differential voltage gain mentioned previously. (The closed-loop gain is lower than this value because of the influence of feedback.)

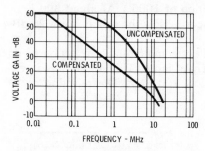

Fig. 4-6. Op amp phase-compensation curves.

The load resistance R_L is also shown although it is assumed in this discussion of the basic equivalents that its value is high in comparison to the output impedance of the operational amplifier. Therefore, it can for the moment be neglected.

The *closed-loop gain* expression is very simple:

$$A_v = \frac{V_{OUT}}{V_{IN}} = \frac{Z_F}{Z_R}$$

Note that the gain of the inverting configuration is entirely dependent on the values of the external feedback components. This is true only if certain requirements are met. The open-loop differential voltage gain must be very high. The intrinsic input impedance Z_i must be much greater than the value of Z_R and the paralleling feedback components. The intrinsic output impedance Z_O must be smaller than the feedback value Z_F.

Another factor is the *loop gain* of the equivalent circuit which amounts to a comparison ratio between the open-loop gain and closed-loop gain.

$$\text{Loop gain} = \frac{A_O}{A_v} = \frac{A_O}{Z_F/Z_R}$$

If we can assume that the intrinsic input impedance is high, the actual input impedance of the equivalent then becomes simply:

$$Z_{IN} \cong Z_R$$

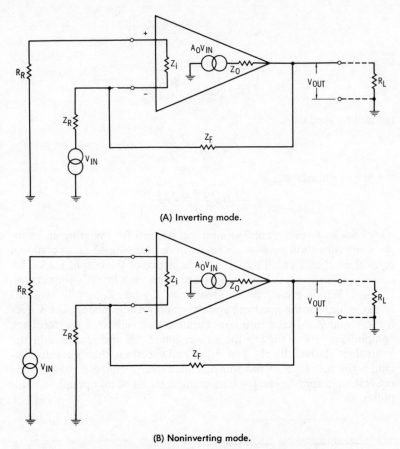

(A) Inverting mode.

(B) Noninverting mode.

Fig. 4-7. Equivalent diagrams of op amp configurations.

The equivalent output impedance is a more complex term because the impedance of the feedback circuit must be considered in conjunction with the intrinsic output impedance of the operational amplifier.

$$Z_{OUT} \cong Z_O \frac{Z_R + Z_F}{A_O Z_R}$$

These requirements are inherent in the design of most operational amplifiers especially those with differential inputs.

Similar operational requirements can be safely assumed for the noninverting configuration, Fig. 4-7B. The basic equations are not identical because the input arrangement differs from the inverting type. Signal is applied to the noninverting side while feedback component must of necessity be applied to the inverting side of the differential input. The basic equations are the closed-loop voltage gain,

115

$$A_v = \frac{V_{OUT}}{V_{IN}} = \frac{Z_R + Z_F}{Z_R}$$

loop gain,

$$\text{Loop gain} = \frac{A_0 Z_R}{Z_R + Z_F}$$

the input impedance,

$$Z_{IN} \cong \frac{A_0 Z_i Z_R}{Z_R + Z_F}$$

the output impedance,

$$Z_{OUT} \cong Z_0 \frac{A_0 (Z_R + Z_F)}{Z_R}$$

A knowledge of the configuration equivalents for inverting and non-inverting operation permits an easier understanding of an operational equivalent circuit which includes the influence of the output load. This basic equivalent applies to both inverting and noninverting operation, Fig. 4-8. Within the equivalent, the terms Z_{IN} and Z_{OUT} will have different values for the inverting and the noninverting modes. The values for Z_{IN} and Z_{OUT} take into consideration the values of the feedback components. Even though the expression Z_R is included within the equivalent dashed block, one must make certain that its value includes the influence of the source impedance. For these reasons, the equivalent is appropriate for both configurations of an operational amplifier.

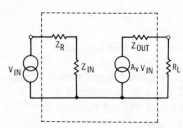

Fig. 4-8. Closed-loop equivalent including output load.

The voltage of the output generator is the product of the closed-loop voltage gain of the operational amplifier times the input voltage, or $A_v V_{IN}$. This voltage divides between the output impedance and the load. Hence, the output voltage itself becomes

$$V_{OUT} = A_v V_{IN} \frac{R_L}{R_L + Z_{OUT}}$$

and stage gain becomes

$$\text{Voltage gain} = \frac{V_{OUT}}{V_{IN}} = \frac{A_v R_L}{R_L + Z_{OUT}}$$

If the ohmic value of the output impedance Z_{OUT} is low in comparison to the ohmic value of the load resistance, there is a further simplification of the equation for voltage gain. For the inverting mode

$$\text{Voltage gain} = A_v \cong \frac{Z_F}{Z_R}$$

and for the noninverting mode.

$$\text{Voltage gain} = A_v \cong \frac{Z_R + Z_F}{Z_R}$$

As stressed previously, it is important to keep the inputs to a differential-input operational amplifier well balanced to reduce ac and dc common-mode components. The dc resistances at both inputs should be kept equal. This does not seem to be the case for the circuit of Fig. 4-2 because resistors R2 and R3 are of different values. However, it must be noted that the feedback resistor R_f is also connected to the inverting input of the amplifier. This has a shunting influence on resistor R3. To compensate for this resistor, R2 must be of lower value and thus, equal bias currents are present at terminals 2 and 3.

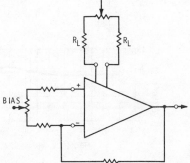

Fig. 4-9. Nulling circuit for correcting dc offset components.

The equalization of differential input-bias currents throughout the amplifier is what establishes proper level shift through the entire amplifier. In many designs, this results in a zero no-signal output. Very fine equalization can also be accomplished by using a balancing potentiometer in the collector circuit of the first differential amplifier (Fig. 4-9). A similar arrangement for nulling can be included in the differential input-bias circuit.

The RCA CA3015 IC, connected as a noninverting video amplifier, is shown in Fig. 4-10A. Note in Fig. 4-10B, that an essentially flat response is obtained up to 25 MHz. A combination of phase-lead and phase-lag compensation permits the bandwidth extension. The Miller-type phase-lag compensation uses a resistor-capacitor feedback arrangement between the collectors of the second differential amplifiers and the collector-base connections between the differential pairs. In Fig. 4-10, these series circuits can be found between terminals 6 and 7, and terminals 11 and 12. These connections produce the compensated response shown in Fig. 4-6. They do not increase the high-frequency response but they do eliminate instability and provide a uniform drop-off of the high-frequency performance.

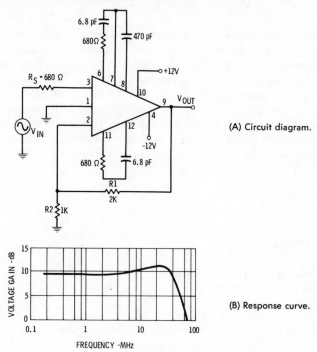

(A) Circuit diagram.

(B) Response curve.

Fig. 4-10. A noninverting video amplifier.

Phase-lead compensation does extend the high-frequency response by providing a low capacitive-reactance path between the base of the driver and the base of the output stage of the operational amplifier for high-frequency signal components. This is done by inserting a capacitor between terminals 7 and 8 (See Fig. 4-1). As shown in Fig. 4-10, the value of this capacitor is 470 pF. This phase-lead high-frequency coupling extends the response above 20 MHz. Above the limit of the flat response, there is a drop-off that corresponds to the 6 dB per octave set by the phase-lag compensation.

A term often referred to in regarding operational amplifier performance is the *slew* or *slewing rate*. As applied to a sine-wave signal,

118

it refers to how well a given amplifier will respond to a high-amplitude, high-frequency signal. The equation is:

$$\text{Slew rate} = \pi f V_{p\text{-}p}$$

where,

Slew rate is in volts per microsecond,
π is the constant 3.1416,
f is the frequency in MHz,
$V_{p\text{-}p}$ is the peak-to-peak sine-wave voltage.

The higher the slew rate of a given amplifier and circuit, the better it is able to produce a full output at a given high frequency. A lower slew rate means that full output cannot be obtained at this frequency.

Phase-lag compensation has a degrading influence on the slewing rate. For this reason, phase compensation, just as much as is needed, is incorporated in the earliest stage possible. Since as the operational amplifier has a relatively high gain after this point, the slewing rate is less influenced than if the phase compensation occurred later in the amplifier.

Three bandpass amplifier arrangements are shown in Figs. 4-11, 4-12, and 4-13. Fig. 4-11 is a simple inverting amplifier and employs

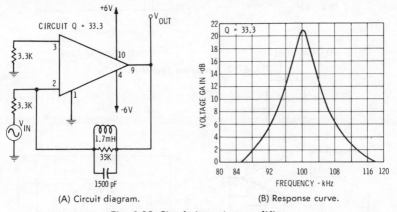

(A) Circuit diagram.   (B) Response curve.

Fig. 4-11. Simple inverting amplifier.

a tuned-feedback link. Inasmuch as it is parallel resonant, there is little feedback at the resonant frequency. However, off resonance, the feedback increases and the gain of the amplifier is correspondingly less. No phase compensating circuits are required for operation at this low frequency.

Fig. 4-12 is unusual. It employs no coils and still has a sharply resonant characteristic at radio frequencies. The twin-T feedback network is a combination of low-pass and high-pass filters. The side containing the two resistors (R1) and capacitor (C2) passes frequencies from the lows up to a certain high-frequency limit. Conversely, the other side with the two series capacitors (C1) and resistor (R2) passes the highs down to a certain low-frequency limit. By proper selection of cutoff frequencies, an intervening range of frequencies is not passed by either filter. Over this range of frequencies, there is no or little feedback and, as a result, maximum amplification. Away from this

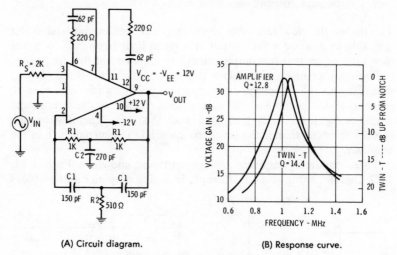

(A) Circuit diagram.

(B) Response curve.

Fig. 4-12. Bandpass amplifier.

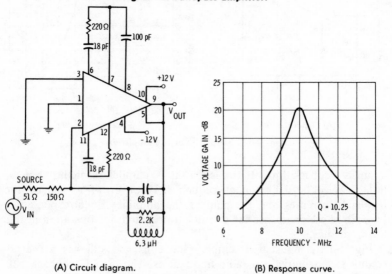

(A) Circuit diagram.

(B) Response curve.

Fig. 4-13. High-frequency amplifier.

120

band, there is great feedback and no amplification. The twin-T network equations showing the relationships between the component are:

$$R1 = 2 R2$$
$$C1 = \frac{1}{2} C2$$
$$F_o = \frac{1}{2 \ (R1) \ (C1)}$$

Fig. 4-13 is a 10-MHz high-frequency amplifier. The feedback combination is a parallel RLC combination tuned to 10 MHz. It minimizes the transfer of a signal of this frequency along the feedback path. Consequently, the amplifier has maximum gain. There is much more feedback off the bandpass and little amplifier gain. Note that the Q is reasonable at this high frequency, and the gain is 20 dB.

OPERATIONAL AMPLIFIERS AT WORK

The versatility and circuit simplicity of operational amplifiers are demonstrated by the following circuits. (The LM number prefix indicates a National Semiconductor Corporation type device; the CA prefix is an RCA type. The basic inverting and noninverting amplifier configurations, as described previously, are represented by the circuits in Figs. 4-14A and 4-14B. Gain of the inverting amplifier is the

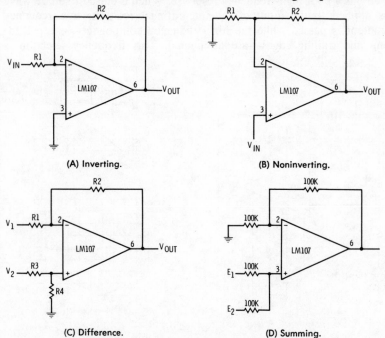

(A) Inverting.   (B) Noninverting.

(C) Difference.   (D) Summing.

Fig. 4-14. Basic operational amplifier applications.

121

quotient of R2/R1; gain of the noninverting version is $(R1 + R2)/R1$.

Difference and summing amplifiers are shown in the circuits of Figs. 4-14C and 4-14D. The two voltages are applied separately to the differential inputs. When resistor R1 is made to equal R3 and resistor R2 is equal to resistor R4 in the circuit of Fig. 4-14C, the difference-amplifier output becomes:

$$V_{OUT} = \frac{R2}{R1}(V_2 - V_1)$$

The parallel combination of R1 and R2 must be made equal to the parallel combination of R3 and R4. In the case of the summing amplifier of Fig. 4-14D, all signals are applied to one of the differential inputs. The output is the sum of the individual applied voltages.

Because the operational amplifier is a dc amplifier, the circuits can be operated with applied dc and/or ac signals. For ac amplification alone, signals to each of the circuits can be applied by way of a coupling capacitor.

Operational amplifiers can be used in a variety of wave-form generating and shaping stages. Typical integrating and differentiating circuits are shown in Figs. 4-15 and 4-16. Basically, an integrating circuit has a poor high-frequency response. When a pulse or square wave is applied to the input of the integrating circuit, its low-frequency content is passed while the high-frequency components—steep leading and trailing edges—are attenuated. This frequency selection is

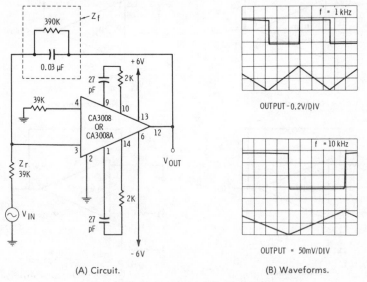

(A) Circuit.   (B) Waveforms.

Fig. 4-15. Using an op amp as an integrator.

done by the characteristics of the feedback network Z_f (Fig. 4-15A). Note that the capacitor passes the high frequencies readily and, therefore, they see less gain. The feedback network has a high impedance to low frequencies, making it relatively inactive. Consequently, the lows pass readily to the output. As a result, the output wave is triangular; it has an exponential rise and fall like the rise and fall of voltages across a capacitor.

Differentiation is a converse type of function (Fig. 4-16). The highs are passed and the lows are attenuated. On application of a pulse or square wave, it is the high-frequency leading and trailing steep edges that are transferred to the output. Long sustained flattops of pulses or square waves do not appear. This portion of the pulse is determined mainly by the low-frequency response and the lows have been attenuated.

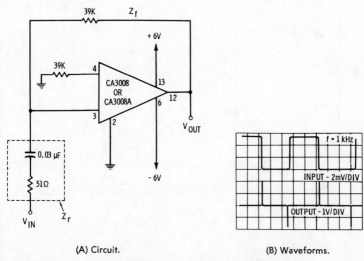

(A) Circuit.   (B) Waveforms.

Fig. 4-16. Using an op amp as a differentiator.

In the circuit shown in Fig. 4-16A, note that the Z_r portion of the feedback combination has a very low impedance except at low frequencies. At low frequencies, the reactance of the capacitor is significant and impedance Z_r becomes high. The influence of this frequency characteristic is shown clearly in the output waveform (Fig. 4-16B) by the sharp spikes in voltage that correspond in time to the leading and trailing edges of the square wave.

Operational amplifiers function well as sine-wave and square-wave oscillators. The example of Fig. 4-17 uses two operational amplifiers to obtain the necessary positive feedback needed to sustain oscilla-

123

tions. In addition, the resistor-capacitor network associated with the first stage acts as a tuned circuit and permits operation only on a frequency determined by its values. That frequency, however, can be tuned over a significant range with the use of potentiometer R3.

Feedback from the output to the input stage is by way of resistor R2. Potentiometer R8 regulates the amplitude of the sine-wave output while the zener diode X1 stabilizes the amplitude of the square-wave output. Note that sine-wave output is obtained at terminal 6 of the first stage. The sine-wave signal is applied to the second stage which acts as a limiter and also causes a resulting square-wave output.

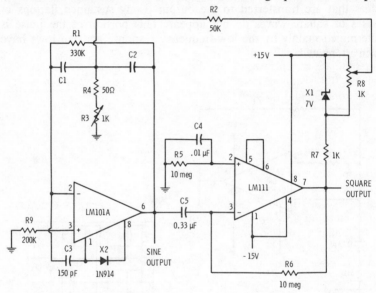

C1, C2	MIN FREQ	MAX FREQ
0.47 µF	18 Hz	80 Hz
0.1 µF	80 Hz	380 Hz
.022 µF	380 Hz	1.7 kHz
.0047 µF	1.7 kHz	8 kHz
.002 µF	4.4 kHz	20 kHz

Fig. 4-17. Sine/square wave generator.

Typical capacitor values for the resistor-capacitor tuning network are shown in the chart. The frequency range over which potentiometer R3 can adjust a specific frequency determined by the RC values, is also shown.

The single, operational-amplifier, sine-wave oscillator shown in Fig. 4-18 uses a Wien-bridge circuit. The constants of the feedback

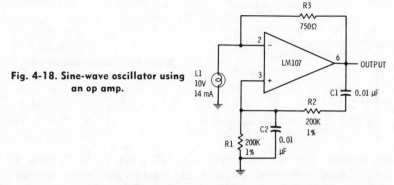

Fig. 4-18. Sine-wave oscillator using an op amp.

network produce a positive feedback of the proper phase for sustaining oscillations only at the frequency determined by the network values. If resistors R1 and R2 are of the same value and so are capacitors C1 and C2, the frequency of operation is:

$$f = \frac{1}{2\pi R_1 C_1}$$

The bridge circuit is stabilized by the small bulb connected between terminal 2 of the differential amplifier input and common. Stabilizing feedback is by way of resistor R_3, while regenerative feedback is obtained by connecting the output back to the noninverting differential input (terminal 3) by way of the frequency-control network.

A similar feedback plan can be used for operating the amplifier as a multivibrator (Fig. 4-19). Frequency is determined by the feedback resistor combination and capacitor C_1. Values given are for oscillation at 100 Hz.

An operational amplifier responds very effectively to the light reaction of a photocell (Fig. 4-20). In such a circuit, there is no need to bias the cell and, therefore, the problem that often arises with cell biasing is avoided. Instead the operational amplifier acts as a very effici-

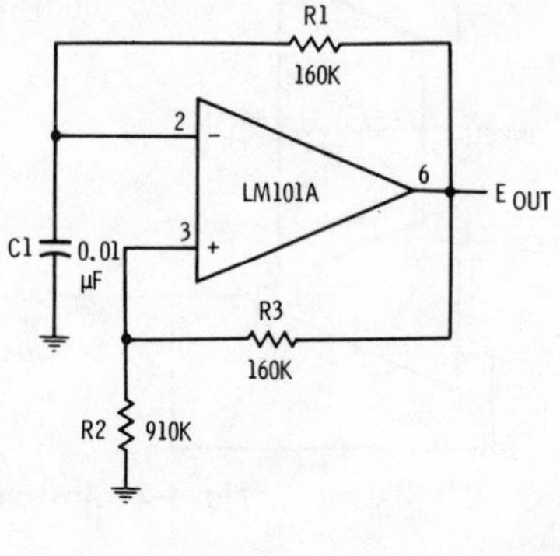

Fig. 4-19. Multivibrator using an op amp.

125

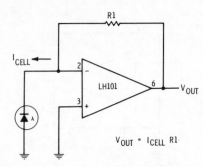

Fig. 4-20. A photocell amplifier.

$$V_{OUT} = I_{CELL} R_1$$

ent current-to-voltage converter. The cell responds to light in the form of a charge motion (current), and since the operational amplifier is current biased, it responds efficiently and quickly, developing a significant output voltage that is a good copy of the light or light variations focused on the cell.

Operational amplifiers are ideal for many types of instrumentation. The example shown in Fig. 4-21 is a three-stage affair that responds to the difference between two applied signals or voltages. The National Semiconductor Corporation's type LM102 operational amplifiers are low input-current devices and are ideal for this type of application. A differential output voltage is obtained and applied to the balanced differential input of the succeeding operational amplifier. The output can then be metered or used in any way desired. With resistors R4 and R5, R2 and R3 of equal values, the voltage gain is equal to R4/R2. In the above example, this would amount to a differential gain of 100.

The high-input impedance of the LM102 makes it attractive in a variety of ac applications. A basic ac amplifier with high-impedance

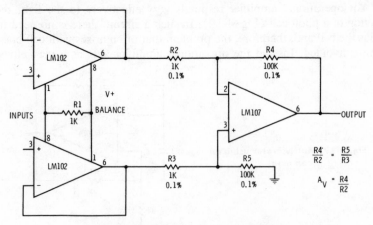

Fig. 4-21. Instrumentation amplifier.

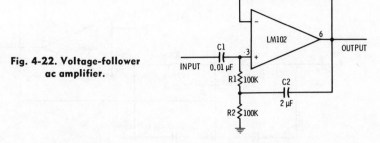

Fig. 4-22. Voltage-follower ac amplifier.

input is shown in Fig. 4-22. It is interesting that for the LM102 op amp, the feedback between the output and the inverting side of the differential input in taken care of internally. Hence, the external connection shown between 6 and minus (−) has already been made. The input resistance for the amplifier is in excess of 10,000 megohms because of the bootstrap connection via capacitor C2. The two 100K resistors in series provided proper dc biasing for the amplifier.

The device also performs equally well as either a low-pass or high-pass active filter (Fig. 4-23). These are simple forms of active filters and require little space and few external components. The basic equations are:

$$C1 = \frac{R1 + R2}{1.414\omega_oR1R2}$$

$$C2 = \frac{1.414}{\omega_o(R1 + R2)}$$

These equations apply to both the low-pass and high-pass configurations. In changing from one type to the other, it is necessary only to switch the relative positions of capacitors and resistors. In the case of the low-pass filter, the feedback path passes the highs more readily

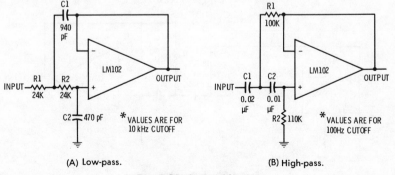

(A) Low-pass.   (B) High-pass.

Fig. 4-23. Active filters.

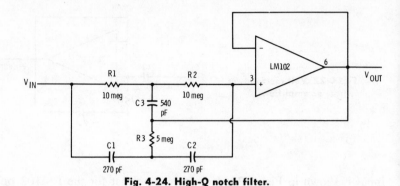

Fig. 4-24. High-Q notch filter.

than the lows. Consequently, the amplifier gain is down at high frequencies. The high-pass filter has a converse relation, with the lows seeing the lowest impedance path through the feedback network.

A combination of low- and high-frequency feedback with suitable near-spaced cutoff frequencies can result in a rejection or notch filter (Fig. 4-24). The constants are chosen with the feedback greatest at the notch frequency. Therefore, the output will be very low in comparison to the output levels above and below the notch. The equations for the notch filter are:

$$f_o = \frac{1}{2\pi R1 C1}$$
$$R1 = R2 = 2\,R3$$
$$C1 = C2 = C3/2$$

To emphasize a particular narrow band of frequencies, the two-stage tuned-circuit amplifier of Fig. 4-25 is appropriate. In this arrangement, feedback is the least over the desired frequency range. The tuned frequency equation is:

$$f_o = \frac{1}{2\pi\sqrt{R1 R2 C1 C2}}$$

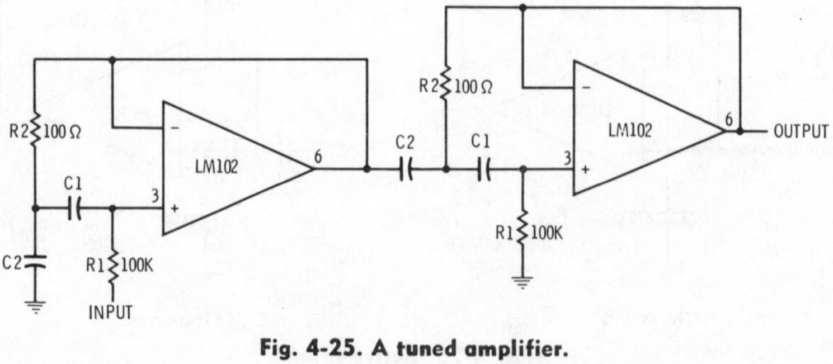

Fig. 4-25. A tuned amplifier.

EXPERIMENT 3: OPERATIONAL AMPLIFIER

General

The operational amplifier is a dc and an ac amplifier of high stability. It can operate with a low gain or a high gain, depending upon the value of three external resistors. In fact, the actual gain of the amplifier depends upon these resistance ratios.

In this experiment you will use the 741 IC operational amplifier, which is available for less than $1.00. It will be connected in various operational modes and will be checked out both as a dc and an ac amplifier.

The two common input arrangements are shown in Fig. 4-26. As a noninverting amplifier, Fig. 4-26A, an amplified output voltage of

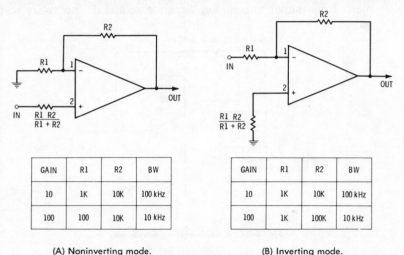

GAIN	R1	R2	BW
10	1K	10K	100 kHz
100	100	10K	10 kHz

(A) Noninverting mode.

GAIN	R1	R2	BW
10	1K	10K	100 kHz
100	1K	100K	10 kHz

(B) Inverting mode.

Fig. 4-26. Input configurations for an op amp.

the same polarity is obtained. Recall in your study of operational amplifier operation that balanced operation is obtained when the effective resistances at each input are identical. In the calculation of the input resistance, it is always necessary to consider the influence of the feedback resistance. In the case of the noninverting amplifier, the input 1 resistance is equal to R1. However, the signal input resistance (2) has an ohmic value which must match the parallel combination of resistors R1 and R2. Typical gain and resistance values are listed in the chart of Fig. 4-26A.

To obtain an inverted output as in Fig. 4-26B, the signal is applied to the same input as the feedback component. Again, the resistance values determine the stage gain. Stage gain approximates the ratio of the feedback resistance R2 over the input resistance R1. This can be verified from the chart by dividing the ohmic value of resistor R1 into that of resistor R2.

The operational amplifier can be built around either the second or third eight-pin socket on your vector board. The circuits require operation with a split-voltage supply. This can be obtained from two 6-volt lantern batteries. The batteries can be used again for experiments four and five and in the projects included in the last four chapters of the book.

Procedure 1: Noninverting Amplifier

1. Connect the circuit of Fig. 4-27. Be certain to observe the polarities of the 6-volt batteries connected to terminals 4 and 7 of the 741 operational amplifier. In the initial procedure, connect a dc voltmeter from output to common. Separate ac and dc inputs are required.

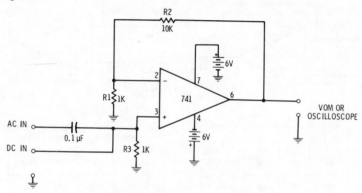

Fig. 4-27. A noninverting amplifier.

2. After construction, turn on the amplifier. Note that the dc voltage measured at the output is very low (usually considerably less than 0.1 volt). In a critical application, it is possible to obtain a precise balance by using appropriate circuits connected to the offset-null pins 1 and 5 of the 741. In our experiment no connections will be made to these pins.
3. The stage has been connected as a noninverting amplifier. Therefore, a + battery voltage applied at terminal 3 will result in a positive voltage at terminal 6. Connect a 1½-volt battery between input terminal 3 and common (positive side of battery attached to terminal 3. A positive dc potential of several volts should be read on the voltmeter connected to terminal 6.
4. Disconnect the battery. Reverse the polarization of the voltmeter connected between terminal 6 and common. Now attach the negative side of the 1½-volt battery to terminal 2; the positive side, to

130

common. The dc output voltage is now negative by a like amount. The zero-output potential of the operational amplifier with no input signal has been displayed. Also, you have seen that an amplified dc voltage appears at the output with the same polarity as the dc voltage applied to the input.

5. Disconnect the input-bias battery and the dc voltmeter. Connect an oscilloscope across the output and an audio generator to the input. The audio signal is applied through the $0.1\mu F$ coupling capacitor.

6. Set the audio oscillator to 1000 hertz. Adjust it slowly, increasing the output to a level that begins to distort the output waveform. Now, decrease the adjustment to the highest nondistorted level. Adjust the oscilloscope vertical-gain control until the pattern occupies approximately ten divisions on the oscilloscope screen. Transfer the oscilloscope to the input and observe how many divisions are occupied by the input sine wave. Output divided by input equals the voltage gain of the operational amplifier. It should approximate ten (10 divisions divided by 1).

7. Momentarily place a second 10K resistor in parallel across resistor R2. What happens to amplifier gain? The gain falls off because it is a function of the R2/R1 ratio.

8. Substitute a 100K resistor in place of the 10K resistor, R2. There is a substantial increase in gain. However, gain and bandwidth are exchanged. The higher the stage gain, the lower becomes the high-frequency limit of the amplifier.

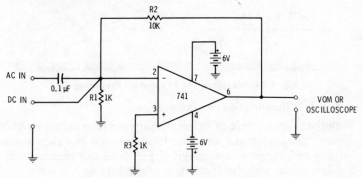

Fig. 4-28. An inverting amplifier.

Procedure 2: Inverting Amplifier

1. Connect the circuit of Fig. 4-28. Now the input on the feedback side is active; the other input is not used but is balanced to common through resistor R3. Connect the voltmeter across the output.

2. Turn on the stage and note the dc output of near-zero volts. Connect the 1½-volt battery to the input, attaching its negative side to terminal 2 and the positive side to common. Note that the output

dc voltage is positive, indicating that the amplifier is operating in its inverting mode.

Reverse the polarity of the input battery and the polarization of the dc output meter. The output shifts negative when input terminal 2 is biased positive.

3. Disconnect the battery and voltmeter. Connect an oscilloscope across the output and an audio generator to the ac input. Adjust the level of the audio signal to just below the point of distortion. Determine the amplifier gain.

4. Shunt a second 10K resistor across the feedback resistor R2. Note the drop in the output and gain, showing again that gain is a function of the ratio of the two external resistors, R2/R1.

Substitute a 100K feedback resistor for resistor R2 to demonstrate the rise in gain with an increase in the ratio.

stratethe rise in gain with an increase in the ratio.

Restore the 10K feedback resistor.

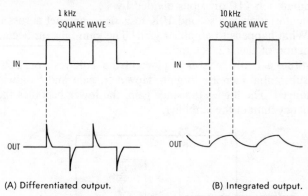

(A) Differentiated output.　　　　(B) Integrated output.

Fig. 4-29. Waveforms of Experiment Procedure 3.

Procedure 3: Differentiation and Integration

1. With the proper choice of external components to match input frequency, the operational amplifier functions well as either a differentiator or integrator. The circuit of Fig. 4-28 will perform well as a differentiator with the application of a 1 kHz square wave.

2. Apply a 1 kHz sine wave to the circuit of Fig. 4-28. Adjust the input level to just below the point of output distortion. Now decrease the level to exactly half. Change the generator from sinewave to square-wave operation and apply a square wave of the same peak amplitude to the circuit. Note the resulting output differentiated spikes (Fig. 4-29A), indicating the emphasis on the high-frequency components in the pulse edges. Poor low-frequency performance results in the falling away of the flat regions of the square wave.

3. Integration can be demonstrated by applying a much higher-frequency square wave. Apply a 10 kHz sine wave to the circuit of Fig. 4-28. Adjust the level to just below the distortion point. Change the generator to apply a square wave of the same peak amplitude. Note how the output has integrated; the edges rounding off with the loss of highs (Fig. 4-29B). Low-frequency components of the original square wave are passed to the output.

5

Multipurpose and Special ICs

A host of integrated circuits have been developed for special applications. Others have been made available in various combinations of connected and unconnected diodes and transistors that with suitable external wiring can perform a variety of circuit functions. This chapter discusses a number of these devices; such as balanced modulators and demodulators, video amplifiers, radio-frequency amplifiers, multipurpose amplifiers, special arrays, and power-circuit voltage regulators.

DOUBLE-BALANCED MODULATORS

Monolithic balanced modulators are exceptionally well-balanced devices that take advantage of the uniformity of monolithic construction. They perform well in the generation of double-sideband and single-sideband signals. They are excellent balanced mixers and phase detectors and can be wired for the demodulation of sideband, double-sideband, fm, and a-m signals.

Ring Modulator

A simplified schematic of the most popular version of a monolithic balanced modulator is given in Fig. 5-1. Transistors are used instead of diodes because they can be produced with uniform characteristics using the monolithic technique. The sequence of operation, however, is similar to the usual diode-ring modulator (Fig. 5-2). In the diode circuit, the carrier strength is great enough to switch the diodes on and off. On the positive alternation of the carrier wave, diodes X1 and X2 conduct. When properly balanced, however, the diode currents in the load cancel. The negative alternation of the carrier wave turns on

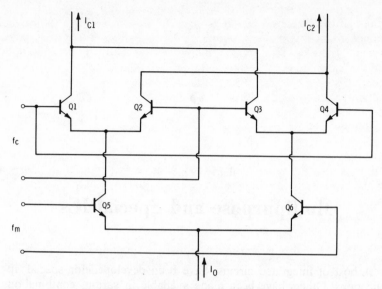

Fig. 5-1. Basic IC double-balanced modulator.

diodes X3 and X4. Again, the balanced configuration results in a net current of zero in the load. There is no carrier output.

The presence of the modulating wave across the opposite corners of the diode ring unbalances the diode currents. In fact, the absolute value of the current drawn by the pair of turned-on diodes depends on the magnitude and polarity of the modulating wave. The modulating wave is of lower magnitude than the carrier wave and does not switch

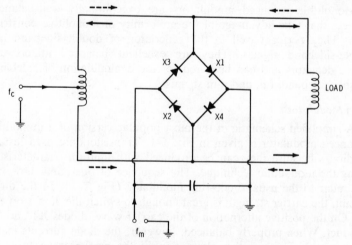

Fig. 5-2. Simplified diode-ring modulator.

any pair of conducting diodes on and off, but only changes the current in accordance with the variations of the modulating wave. It does the same for the opposite pair of diodes during the opposite alternation of the carrier wave. The advantage of the diode-ring modulator, in comparison to the balanced-bridge arrangement, is that the circuit is active for both alternations of the carrier wave, delivering twice as much output.

It is a fact that the modulating wave also sees a balanced circuit and does not appear in the output. Frequencies for which the ring is not balanced are the carrier plus the modulating wave $(f_c + f_m)$ and the carrier minus the modulating wave $(f_c - f_m)$. These are the desired sum and difference frequencies that carry the modulating information; a double-sideband output with suppressed carrier is produced. Such a circuit also produces a series of odd harmonics of the modulating frequency, which are $(f_c \pm 3f_m)$, $(f_c \pm 5f_m)$, etc. When the feedback circuit is suitably arranged, such a balanced circuit is able to emphasize a preferred harmonic in the output. This will be discussed in more detail later.

Balanced Transistors

Balanced transistors provide an ideal double-balanced modulator circuit, Fig. 5-1. A pair of differential amplifiers are used with a pair of transistors that inject the modulating-wave component into the emitter circuits of the differential amplifiers. The carrier wave is applied between bases of the differential pairs as shown. The magnitude of the carrier wave is selected to switch the transistors on and off.

A positive alternation of the carrier wave turns on transistors Q1 and Q4. Inasmuch as the differential transistors are balanced, the carrier currents cancel in the collector output. The negative alternation of the carrier wave switches on transistors Q2 and Q3. This is similar to the switching of the diode pairs in the ring modulator. The two collector currents I_{C1} and I_{C2} are equal, as are the emitter currents. In fact, the constant current I_O is the sum of the two currents. The two collector currents remain equal in value so long as the voltage between the bases of transistors Q5 and Q6 is zero.

When a modulating wave is applied across the bases of transistors Q5 and Q6, the collector currents vary relative to each other, and we have the same effect as the switching modulation of the diodes in the ring modulator. Although the sum of the two currents will always equal the constant current I_O, the relative values will change with the modulation. Unbalancing the ring produces *sum* and *difference* modulation components in the output.

A popular integrated-circuit balanced modulator is the Motorola MC1596G. The internal circuit is shown in Fig. 5-3A. Note the two differential amplifiers at the top with the modulating transistors directly below. A pair of transistors at the bottom serve as the constant-current sources. This low-resistance source is essential for a balanced operation, for an ability to reject common-mode components, and for maximum rejection of carrier and signal components in the output. A diode provides temperature compensation in the constant-current circuit.

A typical circuit is shown in Fig. 5-3B. Note that the carrier is felt between terminals 8 and 7, while the modulating wave is effective between terminals 1 and 4. A balancing carrier-null circuit is connected between the same two terminals. Output is removed between terminals 6 and 9. A push-pull output can be obtained or, by using one of the output terminals relative to common (ground), an unbalanced output can be obtained.

Performance information is given in Figs. 5-3C and 5-3D. A good double-sideband wave with excellent carrier rejection is obtained at the output. Output amplitude as a function of carrier level and modulating-signal input is shown in Fig. 5-3D. A single-supply voltage-source version is given in Fig. 5-4.

A popular double-balanced modulator is the SL-640 (Plessy Semiconductors), which can be adapted to a variety of applications because its frequency range extends between 1 Hz and 70 MHz (Fig.

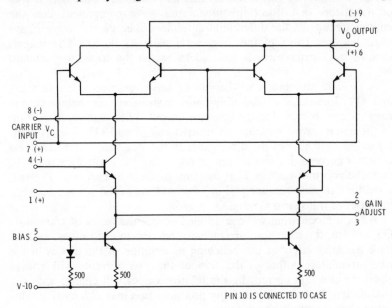

(A) Internal circuit of the MC1596G.

Fig. 5-3. Balanced-modulator

5-5). It operates at reduced performance up to as high as 150 MHz. The device has good carrier and signal rejection. Few external components are needed when it is connected as a balanced modulator.

The carrier and signal null potentiometers are not needed unless the ultimate in suppression is desired. The controls are adjusted alternately. First with the carrier but no signal, potentiometer R1 is adjusted for minimum output. Conversely, with a modulating signal and no carrier, potentiometer R2 is set for minimal leakage to the output. The input impedance on the carrier side is 1000 ohms and 4 picofarads; input impedance on the signal side, 500 ohms and 5 picofarads.

The SL-640 has two outputs. The terminal 5 output has a relatively high impedance of 340 ohms and 8 picofarads. Terminal 6 is an emitter-follower output. This latter output is low and, to avoid distortion, should not be used to drive a capacitive load. It should operate into a resistive load of not less than 560 ohms. In operation and especially at high frequencies, terminal 1, which is connected to the case, should be grounded directly. Terminal 2 should be decoupled to ground through a capacitor which has a very low reactance at the

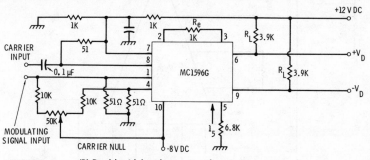

(B) Double-sideband suppressed-carrier generator.

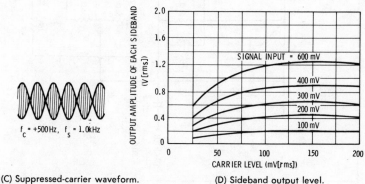

f_c = +500 Hz, f_s = 1.0 kHz

(C) Suppressed-carrier waveform.

(D) Sideband output level.

IC (Motorola MC1596G).

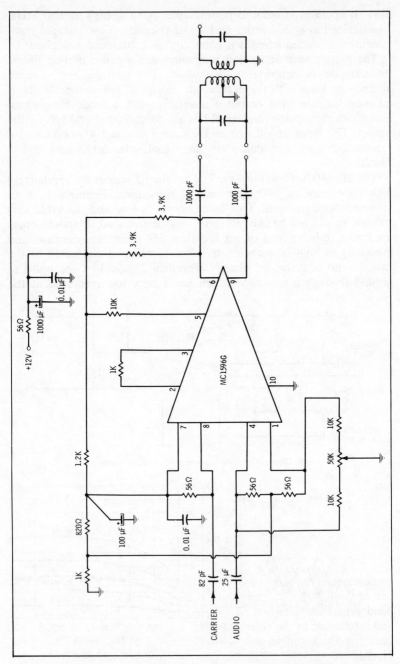

Fig. 5-4. Balanced-modulator circuit using a single supply-voltage source.

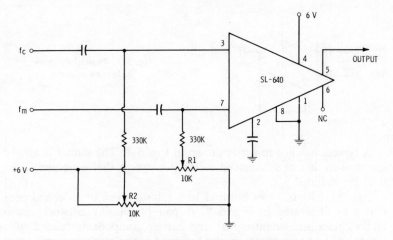

Fig. 5-5. Another balanced-modulator circuit.

carrier and signal frequencies. The system common, terminal 8, is also grounded directly.

Basic Circuits

An advantage of the balanced modulator is that coils and tuned inputs and outputs are not a necessity as compared to the diode-ring

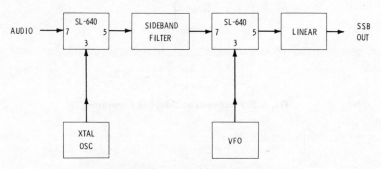

Fig. 5-6. Simple single-sideband generator.

modulators. The functional block diagram of Fig. 5-6 shows how two of these balanced modulators can be used in a simple single-sideband generator. The first SL-640 operates as a balanced modulator and is followed by the sideband filter. The output of the sideband filter goes directly to the second SL-640 which operates as a mixer, stepping the sideband signal up or down in frequency as the need may be. The associated oscillator can be a vfo.

141

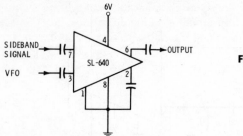

Fig. 5-7. Typical double-balanced mixer.

A typical balance mixer is shown in Fig. 5-7. The circuit is almost identical to that of a balanced modulator except that no nulling controls are included.

Two SL-640 ICs can be used in a phasing-type of sideband generator as illustrated in Fig. 5-8. A pair of phasing networks form audio components related 90° and carrier components related 90°. Sidebands add or subtract, in the balanced-modulator common output circuit.

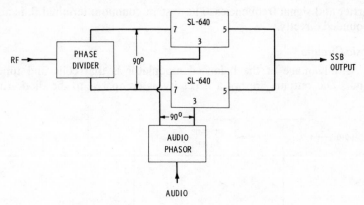

Fig. 5-8. Phasing-type sideband generator.

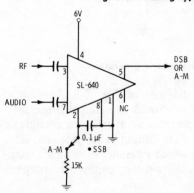

Fig. 5-9. An a-m modulator.

142

The balanced modulator can be used for other communications applications, including that of an amplitude modulator (Fig. 5-9). Note that the connection is the same as for a balanced modulator except for the nulling circuits. In fact, the only change is the addition of the 15K resistor between terminal 2 and ground. This resistor allows car-

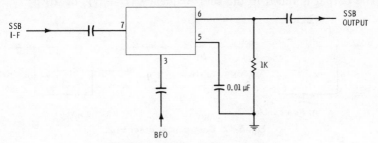

Fig. 5-10. A single-sideband detector.

rier leakage to the output, producing the a-m signal. Multiple switching arrangement, as shown, can be used to switch the resistor in and out of the circuit, making an easy changeover between double-sideband output and a-m output.

It is expected that a double-balanced circuit of this type can also be used as a single-sideband detector (Fig. 5-10). In this arrangement, the *difference* audio component is removed at terminal 6. Terminal 5 is decoupled to ground and assists in removing any *sum* frequency component that develops in the demodulation process.

Balanced modulators are also well adapted for use as multipliers or dividers. In these applications, an output resonant circuit must be employed, emphasizing the desired output frequency and attenuating fundamentals and other harmonic components. As shown in Fig. 5-11A, for the usual application of the double-balanced modulator, there are signal and carrier inputs and a single output signal. The two input frequencies mix, cancel each other, and produce the sum $(f_1 + f_2)$ and difference $(f_1 - f_2)$ components in the output (Fig. 5-11B).

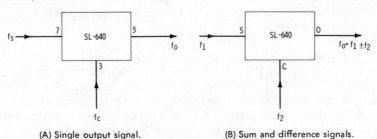

(A) Single output signal.   (B) Sum and difference signals.

Fig. 5-11. A balanced mixer.

When the device is used as a multiplier or divider, only a single input signal is required, Figs. 5-12 and 5-13. Either the input signal is applied to both inputs or a second input is derived from the output. In the doubler circuit of Fig. 5-12A, the input signal is applied to both the signal and carrier inputs, terminals 7 and 3 of the device. The output is tuned to the sum frequency $(f_1 + f_1)$, or $2f_1$.

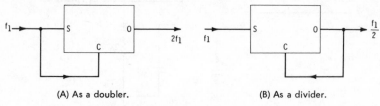

(A) As a doubler.   (B) As a divider.

Fig. 5-12. Diagrams of other uses.

In Fig. 5-12B, the signal is applied to just the signal input of the device. However, the carrier and the output terminals are tied together. The output is tuned to the difference frequency, therefore:

$$f_0 = f_1 - f_0$$
$$2f_0 = f_1$$
$$f_0 = \frac{f_1}{2}$$

In this manner, the output frequency is made to be one-half of the input frequency.

Cascaded ICs

Two doublers can be connected in cascade to obtain a multiplication of four as shown in Fig. 5-13A. In this case, the output frequency is four times the input frequency.

Fig. 5-13B shows a tripler connection. In this case, the first device acts as a doubler to produce an output frequency of $2f_1$. This frequency is, in turn, applied to the input of the second device. However, the carrier input of the second device is the original frequency f_1. Now by tuning the output to the sum frequency, an output signal is developed which has a frequency of 3 times the input frequency:

$$f_0 = 2f_1 + f_1 = 3f_1$$

The same technique can be used to obtain frequency divisions by 4 or 3 as shown in Figs. 5-13C and 5-13D, respectively. In Fig. 5-13C, the input device operates as a two-to-one divider followed by a second two-to-one divider, producing an output frequency of $f_1/4$.

In the three-to-one divider, the input device acts as a divider while the output device functions as a multiplier. The multiplier is needed to obtain the signal of appropriate frequency for the carrier input of the first device. This frequency is $2f_1/3$. When the output of the first device is tuned to a difference frequency, one obtains:

$$f_0 = \frac{2f_1}{3} - f_1 = \frac{f_1}{3}$$

When multiplied by two by the second device, this component ends up as a frequency of $2f_1/3$ at the output of the second device. Additional balanced modulators can be added to obtain various integral and fractional multiplications and divisions.

VIDEO AMPLIFIERS

The application of video amplifiers is not confined to television. They are used extensively in other types of electronic systems that

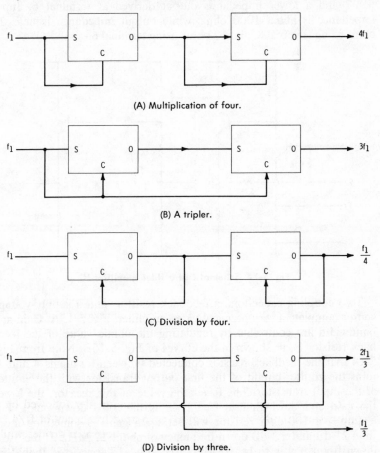

(A) Multiplication of four.

(B) A tripler.

(C) Division by four.

(D) Division by three.

Fig. 5-13. Multiplier and divider systems using two cascaded SL-640 ICs.

employ switching and other nonsinusoidal waveforms. Integrated-circuit video amplifiers are found in low-pass, high-pass, and band-pass amplifier chains as well. Frequency response can be extended upward into the tens of megahertz (without using the old familiar peaking coils seen so often in discrete-component type video amplifiers).

Integrated-circuit video amplifiers have great versatility. Terminals are brought out that can be used to establish very special amplifier characteristics. An example is the Signetics Corporation SE501, which has adjustable gain and impedance characteristics and great versatility (Fig. 5-14). The input system consists of two cascaded common-emitter amplifiers; the output system has two emitter-follower possibilities, one with a lower impedance output than the other. The output can be removed at terminal 8, or terminal 8 can be joined with terminal 7 and a lower impedance output derived at terminal 6. Input impedance is about 1000 ohms while output impedance is near 25 ohms using terminal 8, and 12 ohms using terminal 6.

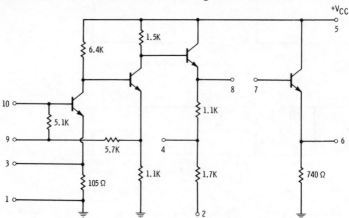

Fig. 5-14. Schematic of a video amplifier IC.

Two coupling capacitors, a feedback resistor, and a supply-voltage source complete a very practical video amplifier, Fig. 5-15A. Gain and bandwidth are controlled by regulating the ohmic value of the feedback resistor R, as shown in the curves of Fig. 5-15B. Note from Fig. 5-14 that the feedback resistor connected between terminals 4 and 3, joins the emitter circuit of the first output transistor with the emitter of the input transistor. The lower the value of this resistor, the lower the gain and the higher in frequency is the 3-dB down point. For example, a 100-ohm resistor will set up a gain of something more than 21 dB and a 3-dB down frequency of about 27 MHz. If the 3-dB down frequency is to be 10 MHz, the size of the resistor should be about 2000 ohms. Under this condition, amplifier gain will be about

27 dB. The values of the indicated parts set up a low-frequency point of about 300 Hz.

Gains in excess of 40 dB can be obtained by cutting back drastically on the negative feedback (Fig. 5-16). Note that terminals 1 and 2 are connected together and grounded. The 3-dB frequency of this connection is only 2.5 MHz; input impedance is 80 ohms and the output impedance is 60 ohms.

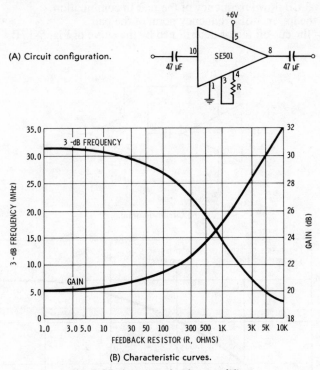

(A) Circuit configuration.

(B) Characteristic curves.

Fig. 5-15. A practical video amplifier.

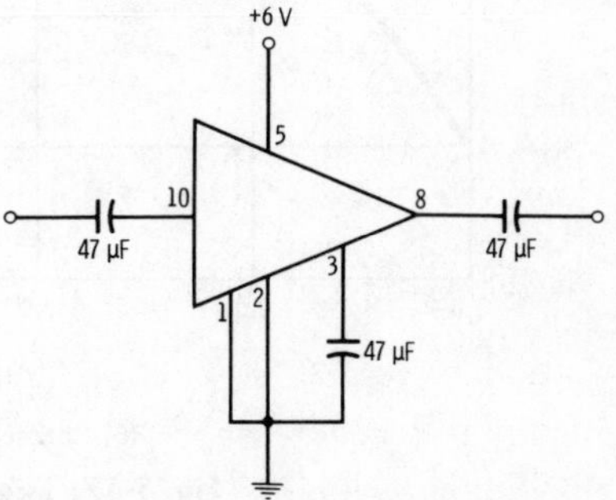

Fig. 5-16. Connecting an IC for high gain.

Power gain or a high gain-bandwidth product can be obtained by using these devices in cascade as illustrated in Fig. 5-17A. Recall that when cascading amplifiers, the low-frequency gain is the sum of the individual-stage dB gains. The 3-dB down frequency of the combination has a value of:

$$f_h = Xf$$

where,

f_h = 3-dB down frequency of the pair in combination,
f = the lower 3-dB frequency point of the pair,
X = the cut-off factor determined by the curve of Fig. 5-17B.

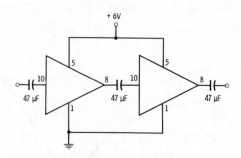

(A) Circuit diagram.

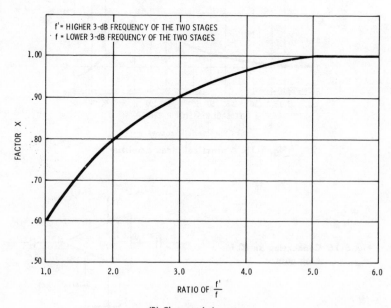

(B) Characteristic curve.

Fig. 5-17. Two-stage amplifier.

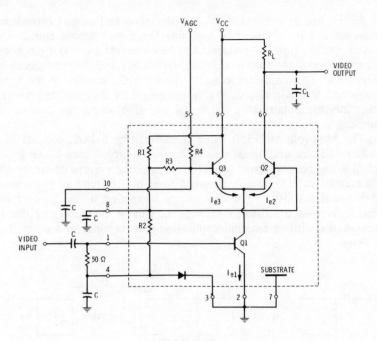

(A) Schematic.

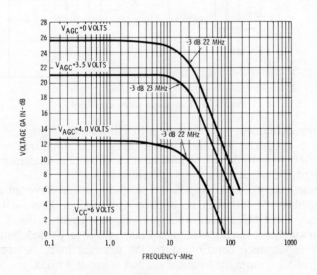

(B) Response curve.

Fig. 5-18. Video amplifier using a Motorola MC1550 IC.

In the use of this video IC, proper input and output impedance matches can be obtained by selecting the most suitable emitter-follower output. Input impedance can be increased by making a bootstrap capacitor connection between terminals 3 and 9 in the circuit of Fig. 5-14. This capacitor connects between the emitter of the input stage and the base circuit, taking advantage of the impedance-rising characteristic obtained by the signal variation across the emitter resistor.

The Motorola MC1550 video amplifier, Fig. 5-18A, consists of a differential pair and the associated current source. Video signal is applied to the base of transistor Q1 and varies the current drawn by the differential pair. Output is removed from the collector of the second differential transistor. The agc voltage is applied to the base of the first differential transistor, regulating the gain of the video amplifier in accordance with an external application that responds to a dc control voltage.

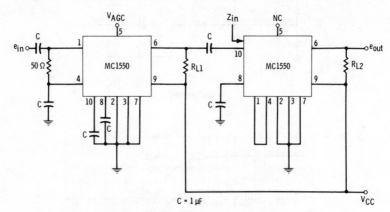

Fig. 5-19. Two-stage video amplifier.

Gain and frequency response as a function of agc voltage is shown in Fig. 5-18B. The top curve would indicate performance with no connection made to the agc terminal 5. Curves assume an external load resistor of 750 ohms and a supply potential of 6 volts.

A practical two-stage amplifier, Fig. 5-19, would have an overall gain of about 55 dB with no applied agc voltage. The 3-dB down frequency is 6.4 MHz as shown by the response curve in Fig. 5-20. In such an amplifier, the high-frequency response is very much influenced by any capacitance contributed by the external load. A capacitive loading of 5 pF starts the frequency rolloff at about 9 MHz.

The first stage of Fig. 5-19 operates as per the external circuit arrangement shown in Fig. 5-18. Load resistor R_{L1} is 1700 ohms. To prevent the second stage from loading down the first stage with a low resistance, the signal is applied to terminal 10. In so doing, the first differential transistor operates as an emitter-follower and has a high input resistance. The output is removed from the collector of the second differential transistor; thus, the pair operates as an emitter-follower common-base amplifier. In this mode of operation, the base input of the source transistor is shorted by connecting a jumper between terminals 1 and 4 (Fig. 5-19).

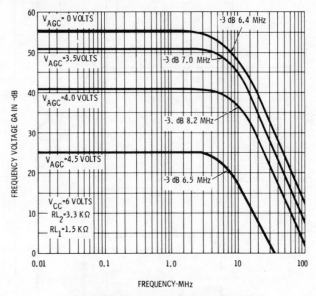

Fig. 5-20. Response curves for Fig. 5-19.

A more elaborate gated video amplifier is the Motorola MC1545 shown in Fig. 5-21. Transistors Q1 and Q4 represent two differential pairs with separate inputs but common outputs. Transistor Q7 is the constant-current source, while transistors Q5 and Q6 split this current between the two differential pairs. When the gate terminal 1 is open, the base of switching transistor Q6 is at a lower potential than the transistor Q5 base. As a result, transistor Q5 is on and Q6 off. Thus, the constant-source current is applied to the Q1-Q2 differential pair. Grounding of the gate terminal or operating at a low potential (less than 0.4 volt) turns transistor Q6 on and Q5 off, and transistor pair Q3-Q4 becomes the operating differential amplifier. In this manner of operation, the video amplifier can be operated as a video switch, making the changeover between two input signals by gating the base voltages of transistors Q5 and Q6.

Any signal applied to the *on* differential pair is amplified and appears differentially at the output. Few external components are required as shown by the simple video amplifier circuit of Fig. 5-22A. Only a single channel is being used in this circuit but it can be switched on and off using a gate pulse. Inputs to the second channel are grounded.

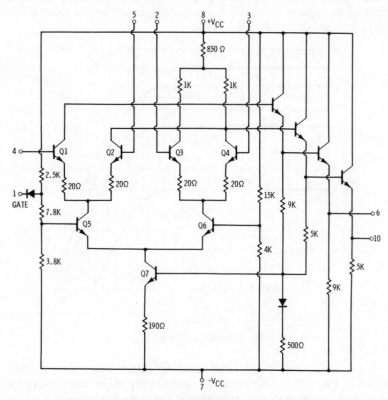

Fig. 5-21. Motorola type MC1545 gated video amplifier.

When switching between two video-signal sources, terminals 2 and 3 would be connected in the same manner as 4 and 8. The frequency response curve is given in Fig. 5-22B. Curves show the excellent high-frequency capability of integrated-circuit amplifiers.

One application of the gated video amplifier is shown in Fig. 5-23. This is a simple amplitude modulator with the modulating wave being applied to the gate input. A potentiometer can control the degree of modulation.

Go a step further and you have a balanced modulator, Fig. 5-24. This is an a-m modulator with carrier suppression. Again the modulating wave is applied to the gate input. Balanced carrier drive is obtained by paralleling the input terminals of each pair and then driving the

152

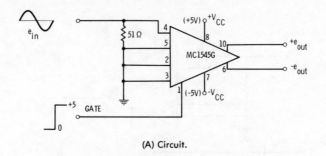

(A) Circuit.

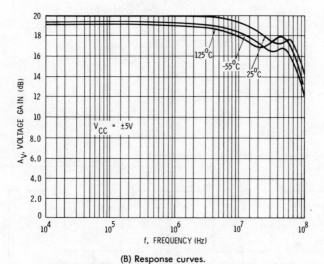

(B) Response curves.

Fig. 5-22. A simple video-amplifier circuit.

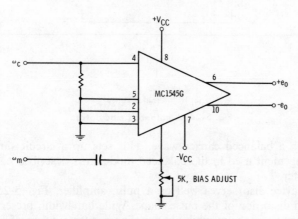

Fig. 5-23. An a-m modulator.

153

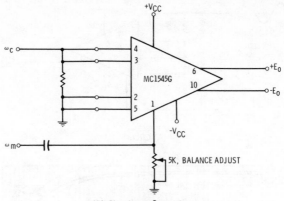

(A) Circuit configuration.

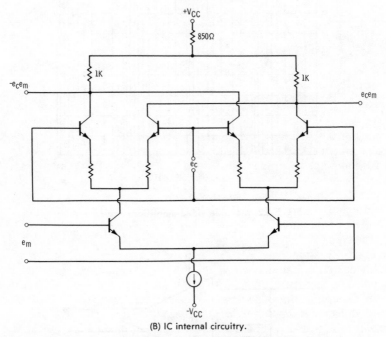

(B) IC internal circuitry.

Fig. 5-24. A balanced modulator.

pairs with a balanced carrier wave. This sets up a circuit similar to the arrangement used in the balanced modulators described earlier in this chapter.

The device also serves well as a pulse amplifier, Fig. 5-25, with minimum distortion of the pulse shape. Wide bandwidth preserves the high-frequency makeup while dc coupling prevents droop of flat sec-

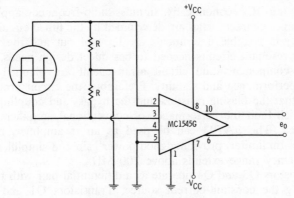

Fig. 5-25. Pulse amplifier.

tions of the pulse. Its differential input and output keeps it free of variations by common-mode components.

RADIO-FREQUENCY APPLICATIONS

The Motorola MC1550 described previously under video amplifiers also performs well in radio-frequency circuits. The source-current transistor is used as the input port, Fig. 5-26A. It is coupled directly to transistor Q3 which functions as a common-base amplifier. The gain of the amplifier can be controlled with a voltage applied to terminal 5. This can be a variable agc voltage, if desired. Split capacitors are used for impedance matching, 50-ohm in and 50-ohm out. Component values for the input and output resonant circuits are shown. Gain figures as a function of agc voltage are given in Fig. 5-26B for 60-MHz operation.

The type MC1550 IC used as a radio-frequency modulator is shown in Fig. 5-27. Component values and rf modulation waveforms for 45-MHz operation are shown. Radio-frequency impedance matching is handled by toroidal coils rather than the split capacitors shown in Fig. 5-26. Otherwise, the radio-frequency sections are identical.

Modulation is applied to the base circuit of transistor Q2 and, as shown in Figs. 5-27B and 5-27C, an excellent modulation characteristic results. Note, too, that the modulating frequency can extend up into the radio-frequency range as well. In fact, the modulating signal can be audio (voice and/or music), video, digital, and other forms of low- and high-frequency information.

Most integrated circuits designed specifically for radio-frequency operation have few components in addition to their transistors, diodes, and perhaps one or two resistors. An example is the Fairchild Semiconductor type μA703, Fig. 5-28. Tuned circuits cannot be designed into ICs economically. In most radio-frequency applications, performance characteristics are determined by the tuned circuits. More versatility is possible if terminals are brought out and the user can plan the resonant circuits needed to best meet design needs. The IC internal components and circuit arrangement are planned for improved performance and stability. Preferably, the circuit arrangement is such that the biasing resistors and the bypass and coupling capacitors, which limit the performance of conventional amplifiers, can be eliminated. The μA703 can be used as an rf amplifier, harmonic mixer, or fm limiter, providing good power gain and simplified tuning. Its frequency range extends above 200 MHz.

Transistors Q3 and Q4 operate as a differential pair with transistor Q5 being the constant-current source. Transistors Q1 and Q2 are

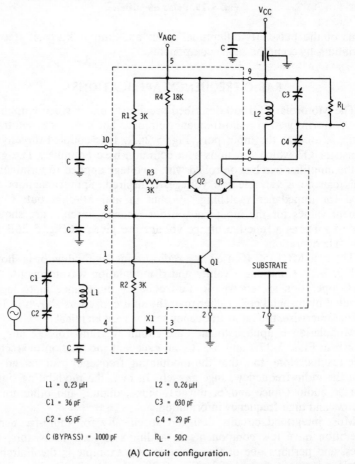

L1 = 0.23 μH	L2 = 0.26 μH
C1 = 36 pF	C3 = 630 pF
C2 = 65 pF	C4 = 29 pF
C (BYPASS) = 1000 pF	R_L = 50Ω

(A) Circuit configuration.

Fig. 5-26. A 60-MHz amplifier

diode-connected and serve as bias source and temperature compensator. Transistors Q4 and Q5 are biased directly by resistors R1 and R2, along with the two diode-connected transistors. Bias for transistor Q3 must be obtained through the external input circuit. (Note that bias diode Q1 must have a dc path between terminals 5 and 3 to the base of transistor Q3.)

Output is taken off between the collectors of the differential pair. Again, the dc collector voltage for transistor Q4 must be by way of the external load.

A 30-MHz amplifier (Fig. 5-29) has a gain of 35 dB and a bandwidth of 1 MHz when using a 12-volt supply. There is transformer matching with toroidal cores at input and output. The input transformer is a one-to-one affair, while the output transformer uses a step-down winding ratio.

The same amplifier can be used as a limiter. Only the turns ratio of the output transformer need be changed to obtain a symmetrical limiting characteristic. The change in reflected load prevents asymmetrical switching of current between transistors Q3 and Q4. In so doing, a swing to saturation is avoided. The condition that must be met is:

$$R_L = 2R2$$

Resistor R2 is an internal one and has a value of 2.5K (Fig. 5-28).

Note that the turns ratio of the output transformer is less in order to reflect a lower-resistance load to the output (Fig. 5-29). Amplifier gain is a bit less and bandwidth somewhat more than in the nonlimiting case.

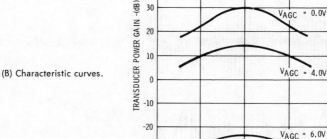

(B) Characteristic curves.

using a Motorola MC1550 IC.

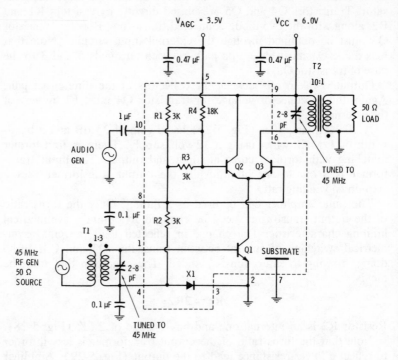

(A) Circuit configuration.

T_1: 6:19 TURNS #32 WIRE ON T12-2 CORE
L_M = 1.1 µH

T_2: 30:3 TURNS #36 WIRE ON T12-2 CORE
L_M = 2.5 µH

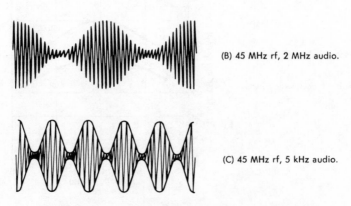

(B) 45 MHz rf, 2 MHz audio.

(C) 45 MHz rf, 5 kHz audio.

Fig. 5-27. Rf modulator using a Motorola type MC1550 IC.

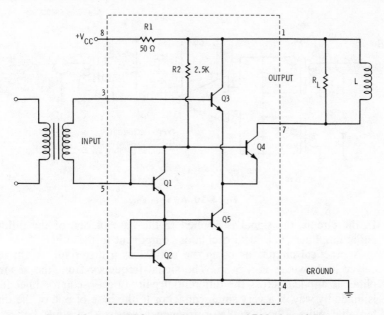

Fig. 5-28. Fairchild Semiconductor type μA703 rf amplifier.

The μA703 device can also be operated as a mixer and performs well as a harmonic mixer. Harmonic mixing is useful for high-frequency mixing systems because of the lower local-oscillator frequency and ease of filtering. In the example of Fig. 5-30, it would be possible to use a 50-MHz local-oscillator source even though the actual mixing frequency is 100 MHz. An incoming signal of 110 MHz could then be converted down to a 10-MHz output difference frequency.

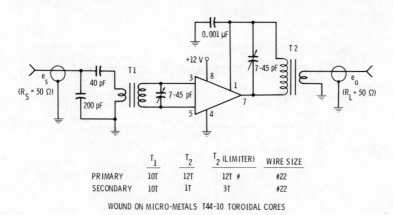

	T_1	T_2	T_2 (LIMITER)	WIRE SIZE
PRIMARY	10T	12T	12T #	#22
SECONDARY	10T	1T	3T	#22

WOUND ON MICRO-METALS T44-10 TOROIDAL CORES

Fig. 5-29. A practical 30-MHz amplifier.

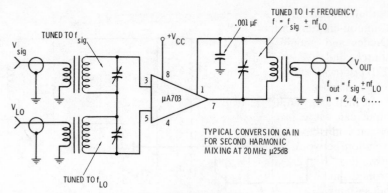

Fig. 5-30. An rf mixer.

In the circuit, the signal is applied to the base of one of the differential amplifiers, the local oscillator component to the other.

A practical circuit for using the μA703 as a mixer for an fm receiver is shown in Fig. 5-31. The signal frequency from the rf amplifier is applied across the differential pair. Local-oscillator injection is made by way of a very small capacitor to the base of one of the differential pairs. Local-oscillator frequency covers the range between 49.35 and 59.35 MHz. The second harmonic component beats with the incoming signal to produce a 10.7-MHz i-f frequency in the output.

A more elaborate integrated circuit, a radio-frequency amplifier, is the Motorola MC1590G, shown in Fig. 5-32A. The input is a common-emitter differential amplifier (transistors Q1 and Q2). They drive a pair of common-base differential amplifiers (transistors Q3 through Q6). Differential outputs are supplied to a four-transistor

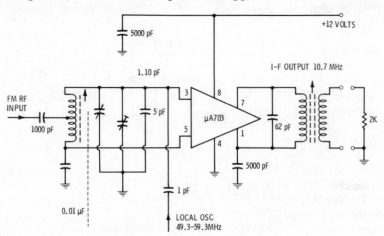

Fig. 5-31. Harmonic mixer stage for an fm receiver.

output configuration, each side consisting of an emitter-follower and common-emitter output transistor (transistors Q7 through Q10). Diodes and unnumbered transistors provide the necessary current sources and stabilized bias system. The accompanying chart (Fig. 5-32B) shows the power gain as a function of frequency, source, and load resistance.

In high-gain amplifiers of this type, feedback between output and input can cause instability and a tendency to self-oscillation. Components must be laid out carefully with short leads and maximum isolation between output and input. Toroidal cores are of particular advantage because of their reduced magnetic coupling. Mount a ground shield between terminals 4 and 8 to isolate the input and output systems.

A practical single-stage tuned amplifier with gains (as a function of frequency) between 30 and 55 dB is shown in Fig. 5-33. The input differential amplifiers are fed with signal from a single-ended source. Capacitive impedance matching is used with data given for 30-, 60-, and 100-MHz operation. The output signal is derived in single-ended fashion also, using capacitors as an impedance-matching means.

Gain, as a function of agc voltage, and the ohmic value of the agc bias resistor R are given in the chart of Fig. 5-33C. Note how the agc action can be controlled with the value of the agc bias and agc resistor. A more gradual gain reduction is obtained with a high agc voltage and a high-value agc resistor.

The RCA CA3040 integrated circuit in Fig. 5-34A has been designed specifically for wideband video and radio-frequency operation up to 100 MHz. The basic configuration is two cascoded amplifier pairs connected in a differential amplifier. Transistors Q1 and Q2 provide a high-resistance input. Their emitters are direct-coupled to the differential input transistors Q3 and Q4. These two transistors, in turn, are cascoded with transistors Q5 and Q6 to form the two cascoded differential pairs. Their outputs drive a pair of emitter-follower output stages. Transistor Q9 is the constant-current source. Assisting in the biasing are the two reference diodes, X1 and X2. The functional block diagram of Fig. 5-34B shows the signal paths.

The device permits biasing by constant-voltage or constant-gain means (Fig. 5-35). In the constant-voltage plan, terminals 7 and 9 are connected together, and the base receives its voltage from the bias network at the junction of resistors R7 and R8 (see Fig. 5-34). No connections are made at terminals 3 and 8. Dc voltage variation is less than 0.1 volt over the temperature range of the device. Possible gain variation is ±2 dB. The constant-voltage characteristic is helpful when there is a dc coupling path to succeeding stages. For many applications, this is the most advantageous arrangement because of less distortion and the ability to handle large signals.

The constant-gain arrangement connects terminals 3 and 9, while terminal 8 is connected to common (substrate). In this connection, there is some dc variation at the output (±0.8 volt), but the ac gain is the same over the entire temperature range.

A typical wideband amplifier circuit is shown in Fig. 5-36. Note that the response is flat up to 20 MHz and the unity gain frequency

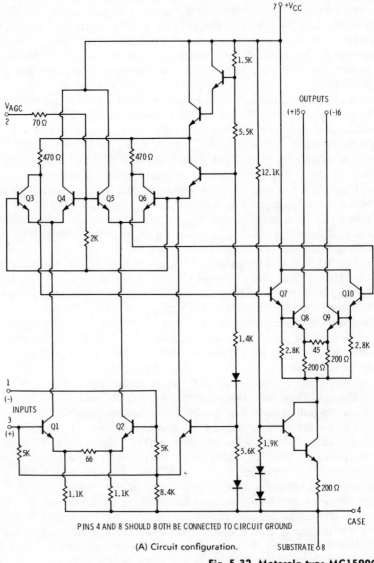

PINS 4 AND 8 SHOULD BOTH BE CONNECTED TO CIRCUIT GROUND

(A) Circuit configuration.

Fig. 5-32. Motorola type MC1590G

162

is in excess of 400 MHz. The circuit uses a single-ended input and a balanced output. No resonant circuits are included and the gain is in excess of 30 dB over the wide frequency range. A small capacitor, C_f, is an output balance control and is adjusted for equal output (at terminals 10 and 11) at the 3-dB down frequency of the amplifier.

Flat response can be extended up to more than 80 MHz by using the simple circuit of Fig. 5-37 which includes a peaking coil in the input circuit.

SPECIAL ARRAYS

Integrated circuits are available with a variety of unconnected diodes, transistors, and larger groupings. Terminals are brought out and can be interconnected according to the desires of the users. These components are fabricated simultaneously on the single silicon chip and have identical characteristics. They respond in the same manner to temperature change.

One such diode array is the RCA CA3019, Fig. 5-38. There are four diodes connected in a bridge or quad, as well as two independent diodes. As covered previously, each diode is really a transistor with collector and base tied together and serving as the anode. The emitter functions as the diode cathode. This method of connection sets up a high-speed diode with the lowest storage time. Recall that this method of connection also avoids the parasitic pnp transistor action to the substrate. There is also a diode possibility between the collector and the supporting substrate. However, that is removed by reverse-biasing the junction by connecting terminal 7 to a dc voltage that is more negative than the diode anode.

There are many applications for such a diode array. An example is the input and output limiter of Fig. 5-39A. By joining terminals 5

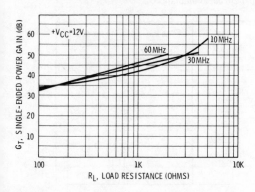

FREQ	SOURCE RESISTANCE
(MHz)	
10	770 Ω
30	600 Ω
60	330 Ω

(B) Characteristic curves.

high-gain rf/i-f amplifier.

163

and 8 and connecting them to ground, the four quad diodes can be correctly connected as a limiter.

Isolated diodes X5 and X6 can operate as a balanced mixer as shown in Fig. 5-39B. Good balance is obtained because the diode characteristics are identical. Conversion gain as a function of oscillator amplitude is shown in the accompanying graph (Fig. 5-39C). Note that good conversion can be attained at vhf frequencies. The same diodes can also be connected into a discriminator circuit.

A balanced modulator is shown in Fig. 5-40. The symmetrical bridge network of identical diodes provides good carrier rejection. Diodes are cut on and off by the radio-frequency carrier. Each acts as a simple diode modulator contributing sideband currents only to the output. An example of the diode-ring modulator (Fig. 5-41) shows how a greater output can be acquired by taking better advantage of both alternations of the carrier. The ring modulator is set up by using all six of the diodes in the array.

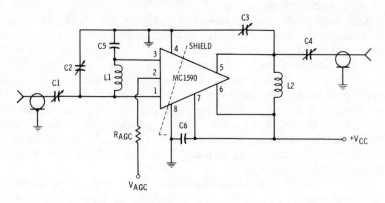

(A) Circuit configuration.

PARAMETER		30 MHz	60 MHz	100 MHz
POWER GAIN (dB)		50.8-54	44.2-46.7	31.6-35.7
BW (MHz)		0.7-1.4	1.9-2.4	7.8-9.2
C1	pF	38	1-30	1-30
C2	pF	1-30	1-30	1-10
C3	pF	1-10	1-30	1-15
C4	pF	1-30	1-10	1-10
C5	μF	0.002	0.001	470
C6	μF	0.002	0.001	470
L1	μH	0.6	0.17	0.07
L2	μH	1.35	0.28	0.13
+V_{CC}	VDC	12	12	12

(B) Parameters per frequency.

Fig. 5-33. Practical

A popular device is the four-transistor array shown in Fig. 5-42. The RCA CA3018 consists of four transistors; two of them with independent terminals for each of their three elements. Another pair of transistors is a Darlington pair with the emitter of one connected directly to the base of the other (transistors Q3 and Q4). No diodes are included in the array. However, the dash line connections of Fig. 5-42 show the collector-to-substrate diode junctions, indicating that the substrate should be connected to the most negative point in the supply-voltage line to make certain that these junctions are always back-biased.

Such triode arrays are useful in the many applications where the user wishes to design his own external circuit arrangement to meet desired specifications. They can be used in i-f, rf, or video amplifiers. The transistors have unusual balance and, therefore, can be connected into differential amplifiers and other balanced configurations.

An example of a video amplifier and its frequency response curve is given in Fig. 5-43. Basically, the amplifier consists of two pairs of common-emitter, emitter-follower combinations. Gain over the flat region of the response curve is near 50 dB.

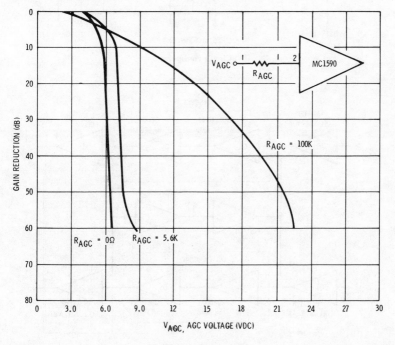

(C) Characteristic curves.

high-gain of amplifier.

165

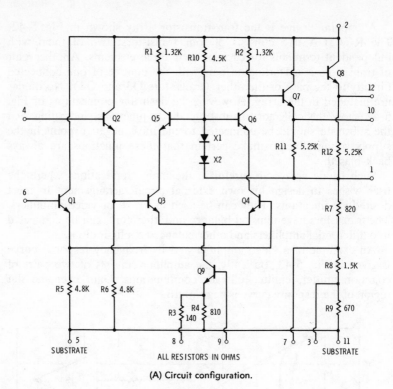

(A) Circuit configuration.

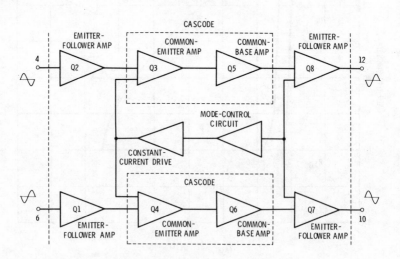

(B) Functional block diagram.

Fig. 5-34. RCA type CA3040 wideband amplifier.

Two feedback loops provide the high dc stability. These paths extend from the emitter of transistor Q3 back to the input, and from the collector of transistor Q4 to the collector of input transistor Q1. The latter connection provides ac feedback over the entire operational range, while the former path provides the low-frequency feedback needed to counteract any low-frequency instability.

Transistor Q1 is in a gain circuit and transistor Q3 places a high load on its output. At the same time, it serves as a low-impedance source for driving the second common-emitter stage, transistor Q4. This connection is taken care of internally as shown in Fig. 5-42. The output of transistor Q4 then drives the emitter-follower output stage, transistor Q2. The output is at low impedance and minimizes the loading influence of any succeeding stage.

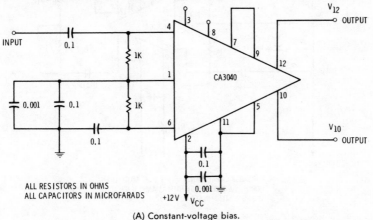

(A) Constant-voltage bias.

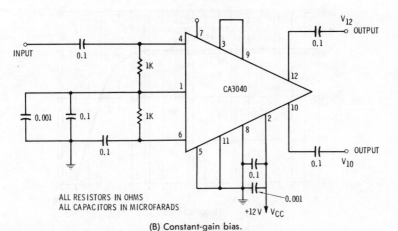

(B) Constant-gain bias.

Fig. 5-35. Biasing modes.

167

The low-frequency characteristics of the amplifiers are determined mainly by capacitors C1, C2, and C3. These can be adjusted for any desired low-frequency cutoff. The 3-dB low-frequency down point for the component values shown is 800 Hz. The upper 3-dB frequency is 32 MHz.

A practical tuned rf amplifier is shown in Fig. 5-44. Transistor Q4 serves as the gain stage. Note that the secondary of the input transformer is connected directly to its base. Transistor Q1 is connected as a temperature-compensating diode. Transistor Q3 is a protective

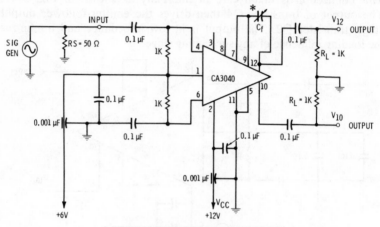

* VARIABLE CAPACITANCE (0.5-1.0 µF) ADJUSTMENT
FOR EQUAL 3-dB BANDWIDTH AT AMPLIFIER
OUTPUTS, TERMINALS 10 AND 12.

(A) Circuit configuration.

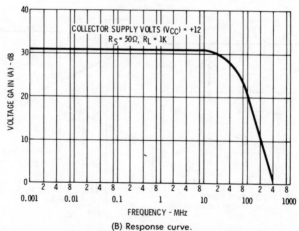

(B) Response curve.

Fig. 5-36. Typical wideband amplifier.

diode, shielding the common-emitter stage from overdrive in the presence of an adjacent transmitter. Transistor Q2 serves as the emitter resistor of the output stage. Agc voltage can be applied here to regulate stage gain. The transformers are wound on high-frequency toroids.

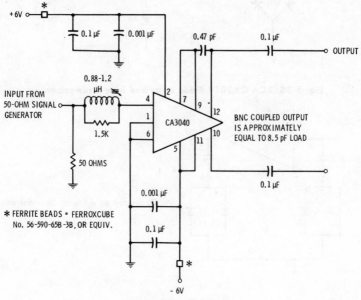

(A) Circuit configuration.

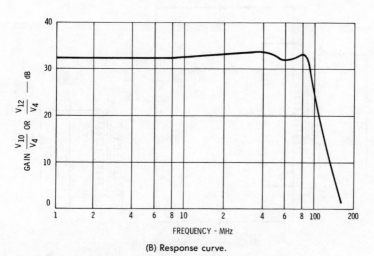

(B) Response curve.

Fig. 5-37. Wideband amplifier with an input peaking coil.

169

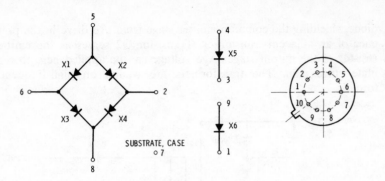

Fig. 5-38. RCA CA3019 diode array and base arrangement.

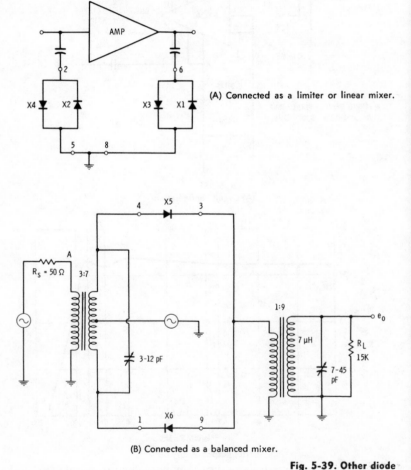

(A) Connected as a limiter or linear mixer.

(B) Connected as a balanced mixer.

Fig. 5-39. Other diode

The circuit of Fig. 5-45 is an example of an untuned final i-f amplifier and second detector. The i-f amplifier gain is approximately 30 dB; transistor Q1 stabilizes the gain with temperature change. Transistor Q2 is the i-f amplifier, while the cascaded emitter-follower combination of transistors Q3 and Q4 function as a detector. The emitter-follower combination presents a high impedance to the collector circuit of the i-f amplifier, transistor Q2. Demodulated audio at low impedance is available at the emitter of transistor Q4. Proper demodulation is obtained because there is an approximate cutoff bias maintained across the emitter junction of transistor Q4.

Even differential amplifiers are available in array form, Fig. 5-46. The RCA CA3049T is a pair of differential amplifiers, each including a constant-current source transistor as well. There is close electrical and thermal matching of the two amplifiers. With suitable external circuitry the device can be made to function up to as high as 500 MHz. Typical power gain at 200 MHz is 23 dB.

The array can be used for a variety of functions including vhf amplifiers and mixers, mixer-oscillator combinations, i-f amplifiers, balanced modulators and demodulators, phase and synchronous detectors, push-pull amplifiers, product detectors, and so forth. Arrays of this type will become more and more common, even in those designs that mainly employ discrete components.

Even more complicated arrays are available. The RCA CA3048 incorporates four elaborate differential amplifiers, two of which are

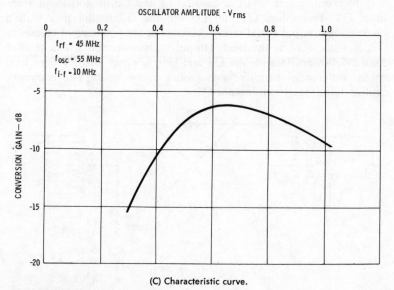

(C) Characteristic curve.

array configurations.

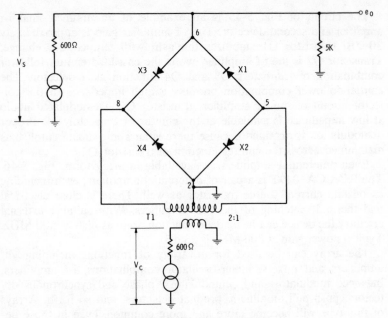

Fig. 5-40. A balanced modulator.

shown in Fig. 5-47. The four amplifiers are identical and include independent inputs and outputs.

A high-impedance input is ensured by the Darlington input transistor Q1. Transistors Q2 and Q3 are the differential pair with a single-ended output signal made available at the collector of transistor Q2. Resistor R17 is the load. Output is direct-coupled to a pair of emitter-followers (transistors Q9 and Q10), which place a light load on the differential amplifier and ensure a high gain. The common-emitter output stage is transistor Q7.

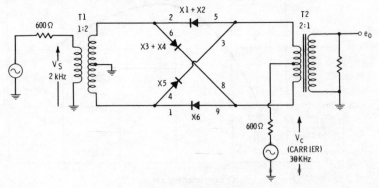

Fig. 5-41. A ring modulator.

A feedback network, consisting of resistors R9, R10, and R6 plus temperature-compensating diode X1, connects the output back to the base of differential transistor Q3. This network ensures stability and biases the output transistor for good dynamic range and freedom from output distortion. Diodes X3 and X4 also provide temperature compensation for both the upper and the lower differential transistors.

The bottom amplifier is identical to the top one. Furthermore, the device includes two more identical amplifiers. The amplifiers have a variety of purposes, related or unrelated. They can be used in various sine-wave and nonsinusoidal oscillators. The circuits can be simple, as shown in Fig. 5-48 for a Hartley sinewave oscillator and an astable multivibrator. Very few external components are needed.

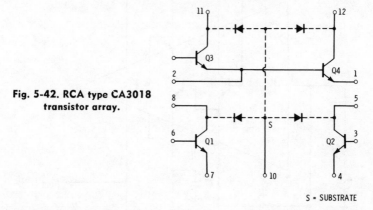

Fig. 5-42. RCA type CA3018 transistor array.

S = SUBSTRATE

A four-channel linear mixer permits the use of all four of the amplifiers (Fig. 5-49). Individual channel gain is 20 dB, assuming a load resistance of 10K or higher.

IC VOLTAGE REGULATION

The integrated circuit provides a compact and exacting control element for a voltage-regulated power supply. The general plan of such a supply, shown in Fig. 5-50, consists of a control circuit that responds to the difference between a component of the load voltage and a reference voltage. The control circuit is then able to regulate the opposition offered by a series resistance. In so doing, the load voltage is held at some desired value.

In integrated-circuit vernacular, as shown in the block diagram in Fig. 5-50B, the functional sections of the regulated supply are referred to as the reference voltage generator, the error amplifier, and the pass device. In practically all integrated-circuit regulator systems, the error amplifier and the reference voltage generator are an inherent

173

part of the IC voltage regulator. According to the current level to be regulated and other factors, the pass device is or is not included in the integrated circuit. Often, it is in the form of one or more discrete power transistors. However, integrated circuits that include their own pass device are available for voltage levels from a fraction of a volt up to 1000 volts. Current capabilities extend upward to $\frac{1}{2}$ ampere or more. External discrete series pass devices can extend current capability to 50 amperes and more.

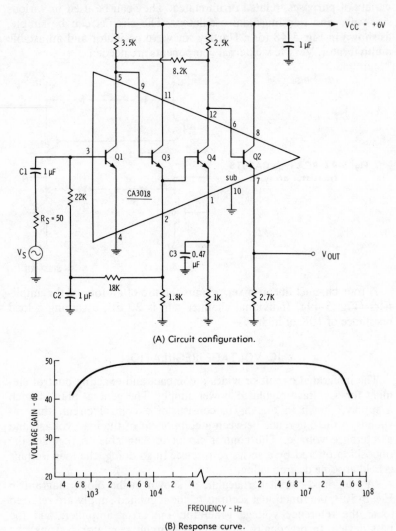

(A) Circuit configuration.

(B) Response curve.

Fig. 5-43. Broadband video amplifier.

Customarily, load regulation in percentage is expressed by the simple equation:

$$VR\% = \frac{V_{No\ Load} - V_{Full\ Load}}{V_{No\ Load}} \times 100$$

In practical applications, the term $V_{No\ Load}$, instead of being the actual no-load voltage, is the output voltage under conditions of minimum load. Often this minimum load is simply the resistive bleeder network connected across the output of the power supply.

The previous expression is accepted as a regulator figure of merit. However, it does not tell the complete story because it does not indicate how well the power supply responds to the rapid changes in loading and load current. Thus, it has been common practice to also indicate how well a regulated supply is able to respond to a fast

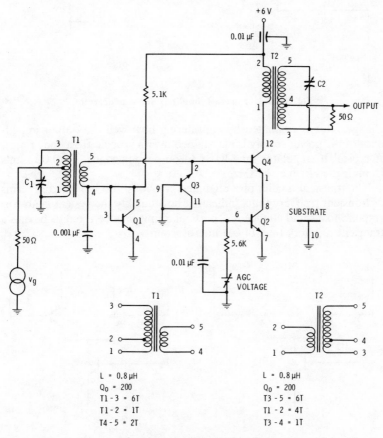

Fig. 5-44. A 15-MHz rf amplifier.

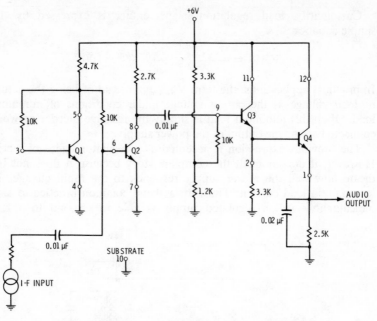

Fig. 5-45. Untuned i-f amplifier and a-m detector.

change. This is evaluated by considering how well the output imped-ance of the power supply is maintained over a specific frequency range. A typical IC regulator may have an output impedance of 0.02 ohms from dc to as high as 1 MHz.

The transient quality of a regulated system is also indicated in terms of transient recovery time, indicating how quickly the system is able to respond to a rapid change. (A particular unit may be rated as having a transient recovery time of 0.3 microsecond.)

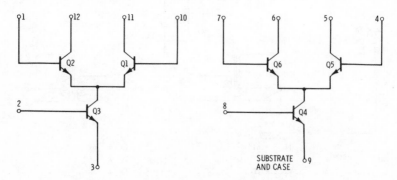

Fig. 5-46. RCA type CA3049T differential amplifier transistor array.

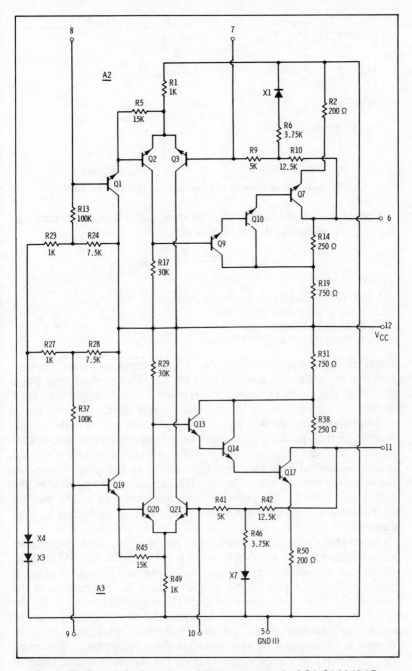

Fig. 5-47. Two of the four ac amplifiers present in the RCA CA3048 IC.

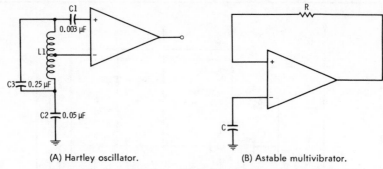

(A) Hartley oscillator.　　　　　　(B) Astable multivibrator.

Fig. 5-48. Circuits using any amplifier of the RCA CA3048.

Output impedance then is an important factor in load-current regulation. It is related to that regulation as follows:

$$LR\% = \frac{\Delta I \times Z_o}{V_o} = 100$$

where,

LR% is the load regulation in percentage,
ΔI is the change in load current,
Z_o is the output impedance,
V_o is the nominal output voltage.

From the equation, it is noted that a low output impedance is preferred in maintaining a constant load voltage. This method furnishes a more exacting figure of merit for a regulated power supply that must respond to high-frequency transients and rapid changes in loading.

Temperature is another factor because heating does have an influence on the operating parameters of a monolithic regulator, particularly if the pass device is a part of the monolithic structure. An increase in loading also increases the current in the pass device, influencing the heating effects in the high-density structure. Briefly, the power that must be dissipated in the regulator varies as the product of the difference between input and output voltage multiplied by load current.

A simplified diagram showing the technique used in many integrated-circuit voltage regulators is given in Fig. 5-51. The old standby differential amplifier is represented by transistors Q3 and Q4. In this stage, a fractional part of the output voltage is compared with the reference voltage. Transistors Q1 and Q2 act as collector loads for the differential pair. Any difference between the two compared voltages results in a change in the collector voltage of transistor Q2. This change is emphasized by two cascaded emitter-followers. This manner of connection places a light load on the output of the differential amplifier.

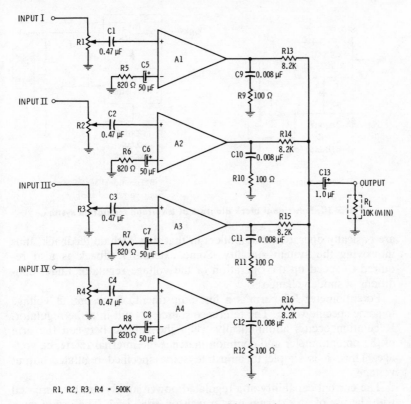

R1, R2, R3, R4 = 500K

Fig. 5-49. Four-channel linear mixer.

Bias at the base of transistor Q6 regulates the resistance of its collector-emitter path which, in turn, is in series with the path between input and output. Transistor Q6 must have sufficient power capability so that it is capable of handling the load current. Since its resistance changes with base bias, any shift in the output voltages away from the desired value produces a corresponding increase or decrease in the resistance of the path and a suppression of the attempted shift in output voltage. The divider network of resistors R1 and R2 determines the output voltage. Their ratio and ohmic values are selected to make their junction voltage a proper value for comparision with the reference voltage.

A basic voltage regulator using an integrated circuit that requires few external components is shown in Fig. 5-52. There are unregulated input and regulated output terminals, a terminal for introducing the control voltage, plus one terminal for setting up a feedback circuit that influences the speed with which the integrated circuit responds to transients and abrupt voltage demands. Many IC voltage regulators

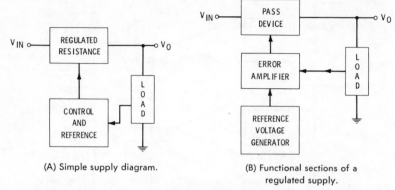

(A) Simple supply diagram.

(B) Functional sections of a regulated supply.

Fig. 5-50. Functional block diagram of a voltage regulation system.

are basically operational amplifiers in that they include feedback, thus improving the system stability. Some external feedback is also required to speed up the operation of the voltage regulator under conditions of sudden change.

Potentiometer R2 permits a fine adjustment of the output voltage to some specific value. The reference source is built into the regulator. Regulation occurs automatically when the voltage between the arm of the potentiometer and common matches the internal reference voltage. Thus, it is simple to adjust to some specified regulated-output voltage.

The current capability of a regulated power supply can be increased with the use of an external pass transistor, Fig. 5-53. The use of such

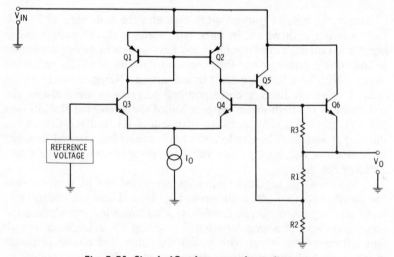

Fig. 5-51. Simple IC voltage-regulator diagram.

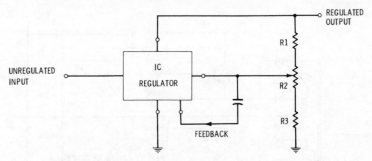

Fig. 5-52. Basic electronic voltage regulator using an IC.

an arrangement with the National Semiconductor LM105 voltage regulator increases its current-handling capability as much as 10 times. Capacitor C1 helps to stabilize the internal reference-voltage generator. Capacitor C3 is needed only when there is a considerable physical separation between the voltage regulator and the power-supply source. The output voltage is determined by resistors R1 and R2, while resistor R3 is used for current limiting, assuring a no-load to full-load regulation of better than 0.1%.

A complete power supply adapted for integrated-circuit application is shown in Fig. 5-54. It employs two National Semiconductor LM100 voltage regulators and supplies a split voltage output (+15 and −15 volts) and a grounded junction. This type of dual regulation is particularly useful in the many types of integrated circuits that provide best balance when operated with a dual power supply (both positive and negative supply voltages). The transformer has dual secondaries, each connected in a full-wave circuit. External pass transistors are used to increase the current capability.

The functional plan of a high-powered voltage regulator shown in Fig. 5-55 is a three-section arrangement. The Motorola MC1560 and MC1561 regulators can provide a ½-ampere load current without the

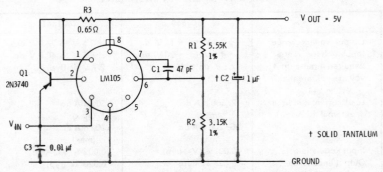

Fig. 5-53. External transistor used as a pass device.

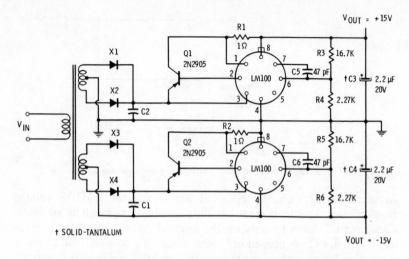

Fig. 5-54. Practical power supply for integrated-circuit application.

addition of any external pass device. They are able to supply an adjustable voltage over a wide voltage range, as well as operate with a substantial range of input voltages. Regulator performance is shown in Table 5-1.

Enclosed within the dashed block at the left (Fig. 5-55) is the reference-voltage generator followed by a dc level-shifting series voltage regulator. The third section is a unity-gain series regulator with a current capability of 500 mA.

The reference-voltage generator consists of a zener diode and an associated constant-current source. The regulated voltage of 3.5 volts is made available at the junction of resistors R_a and R_b. The zener diode and the two diffused resistors have positive-temperature coefficients, while the diode-biased junctions have a like but negative-

Table 5-1. Regulator Performance

	MC1560	MC1561
Output voltage range	+2.5 V to +17 V dc	+2.5 V to 37.0 V dc
Input voltage range	+8.5 V to +20 V dc	+8.5 V to 40.0 V dc
Required input-output voltage differential ($V_{in} - V_{out}$)	2.1 V dc (typ)	2.1 V dc (typ)
Load current (without external transistor)	R: 500 mA dc (max)	500 mA dc (max)
	G: 200 mA dc (max)	200 mA dc (max)
Input Regulation	0.002%/V_{in} (typ)	0.002%/V_{in} (typ)
Output impedance (dc to 100 kHz)	0.025 Ω (typ)	0.020 Ω (typ)

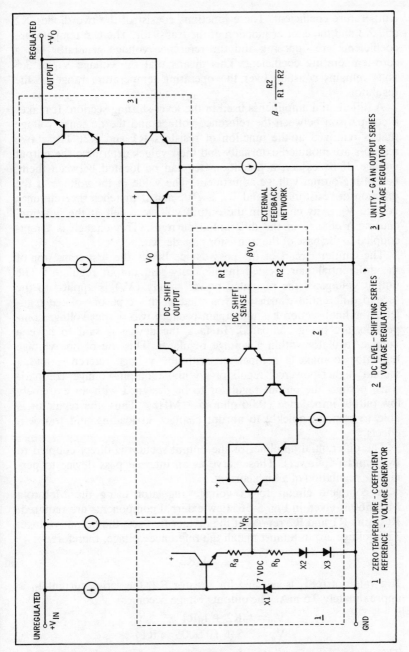

Fig. 5-55. Simplified functional diagram of a Motorola MC1560 voltage regulator.

183

temperature coefficient. These junctions consist of the two diodes X2 and X3 and the emitter junction of the transistor. The two temperature coefficients are opposing and the reference voltage generator has a zero-temperature coefficient. This means that the voltage V_R of 3.5 volts remains constant over the operating temperature range of the regulator.

A differential amplifier is used in the level-shifting section. It makes a comparison between the reference voltage and the dc sense voltage that is removed at the junction of resistors R1 and R2. These two resistors are mounted externally and their values determine the output voltage. If necessary, a potentiometer can be located between them, permitting output-voltage adjustment. The value of the voltage at the junction of resistors R1 and R2 is βV_o and it matches the reference voltage V_R. Any change in the output voltage is felt at the junction, causing a change in the differential currents. This change is direct-coupled to the base of the transistor pass device.

The emitter circuit of the pass device biases the first transistor of the differential pair located in the unity-gain output regulator. The output voltage of the regulated power supply (V_o) is applied to the second differential transistor. Note that in this type of voltage regulator the final section has unity gain because two output-voltage components are being compared. In fact, the device is said to have a voltage regulator within a voltage regulator. This use of one voltage regulator to make a comparison with the voltage reference, and a second higher-powered regulator to make output-voltage comparisons, permits the output regulator to be designed with an extremely low output impedance (0.06 ohm at 1 MHz). Thus, the regulator is able to respond quickly to abrupt changes in loading and transient effects.

The differential amplifier of the output section is direct coupled to two emitter-followers. These serve as an internal pass device to permit the regulation of a high current.

A complete circuit for a voltage regulator using the Motorola MC1560 is given in Fig. 5-56. Few external components are required. Resistors R1 and R2 represent the divider that sets the output voltage. The voltage at pin 8 must match the reference voltage, therefore:

$$V_8 = I_R R2$$

If a value of 6.8K is selected for resistor R2, the series current I_R is approximately 0.5 mA. The output voltage becomes:

$$V_o = I_R R2 + I_R R1$$
$$V_o = 3.5 + (0.0005 \times R1)$$

This equation indicates that the output voltage becomes a linear function of resistor R1 as shown by the curve in Fig. 5-57. The ohmic

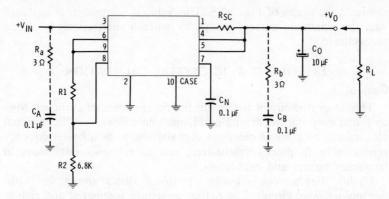

Fig. 5-56. Typical applicatioun of the Motorola MC1560 voltage regulator.

value of resistor R1 for any given output voltage can be selected using this curve. Note that the maximum limit for the type MC1560 is 17 volts, while the type MC1561 is 37 volts. The minimum unregulated input voltage must be greater than 8.5 volts and a minimum of 2.5 volts higher than the desired output voltage V_o.

This voltage regulator is of necessity planned for high-frequency operation because it has been designed to respond well to transients. Any tendency to self-oscillation because of the type of load attached to the input or the output is damped by the series resistor-capacitor combinations found at the input and the output. Capacitor C_N is a noise reduction component that reduces minute low-level high-frequency variations (hash) from the output. Resistor R_{SC} is a component that limits the short-circuit current to a specific value as

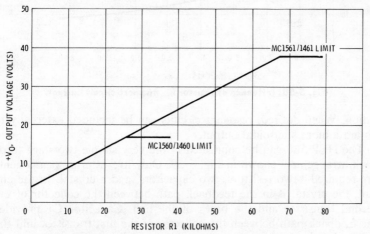

Fig. 5-57. Influence of resistor R1 on output voltage V_o.

shown in the graph of Fig. 5-58. This value is selected in accordance with the maximum current acceptable when a short-circuit is placed across the load.

EXPERIMENT 4: IC CRYSTAL CALIBRATION

General

The integrated circuit has led to the development of compact, low-cost, and easy-starting crystal oscillators and calibrations. Such signal sources can be used as clocks in digital systems, as a base-frequency generator in frequency synthesizer, and as reference oscillators in frequency meters and calibrators.

Usually they are connected in a simple feedback arrangment without any resonant circuit. The output waveform is squared and rich in harmonic content, which is advantageous in crystal-calibrator appli-

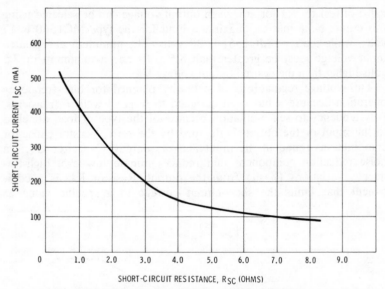

Fig. 5-58. Influence of resistor R_{sc} on short-circuit current.

cation. When desired, resonant circuits can be included externally to obtain a more sinusoidal output.

The HEP 580 can be connected in the very simple two-stage feedback arrangement shown in Fig. 5-59. Only five external components are required—two resistors, two capacitors, and a crystal. In the circuit, the crystal is in the feedback path between the collector of the second transistor and the bases of the first. Capacitor C1 provides the feedback path between the collector of the first transistor and the bases of the second.

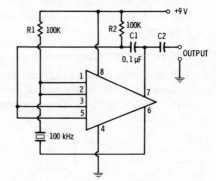

Fig. 5-59. An IC crystal oscillator.

Procedure 1: Crystal Oscillator

1. The oscillator can be constructed permanently around the second 8-pin socket on your vector board. Only two output binding posts are required. Connect an oscilloscope across these terminals.
2. Turn on the oscillator. Collector current should be about 3 milliamperes. Note the square-wave output and the steepness of the sides, suggesting a high harmonic content. The square-wave output is also convenient for making comparisons with other frequency components in phase detectors or as clock pulses for succeeding digital counters.

Procedure 2: Receiving the Calibrator Signal

1. Connect a short length of hookup wire to the output binding post. Wrap the other end of the insulated wire around the antenna staff of a small am/fm radio. Beginning at the 600-kHz low-frequency end of the broadcast band, a strong signal can be picked up every 100 kHz over the a-m dial. The signal is noted by a substantial drop in the background noise level of the receiver at these points. If a local broadcast station operates on one of these frequencies, you will hear a beat-note whistle.
2. Set the receiver on the fm band. The presence of the calibrate signal near the frequency of a local fm station will result in an audible beat note. If the receiver has shortwave bands, the calibrate signal can be heard at close-spaced intervals. The separation between signal components is only 100 kHz, a frequency change that occurs with only a slight movement of the shortwave tuning dial.
3. The calibrate signal can be tuned in more conveniently and with greater accuracy if the receiver has a beat-frequency oscillator. Such oscillators are usually a part of SWL and ham radio receivers. If you have such a receiver, couple the output of the calibrator loosely to the antenna input. Turn on the receiver bfo. You will

note that there is a distinct beat note every 100 kHz over the shortwave and ham bands. This affords an accurate means of calibrating your receiver dial so you can tune it to some precise signal frequency.

Procedure 3: WWV Calibration

1. If you have a receiver that can pick up one of the transmitting frequencies for the WWV signal, you can calibrate your crystal oscillator very accurately. This involves the addition of a small trimmer

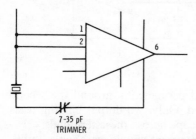

7-35 pF
TRIMMER

Fig. 5-60. Location of trimmer capacitor used in Experiment Procedure 3.

capacitor, connected between the crystal and collector terminal 6, as shown in Fig. 5-60. Make this change.

2. Connect an appropriate antenna to your shortwave receiver and tune in the WWV signal. Couple the output of your crystal oscillator to the antenna input. Usually you only need to wrap a few turns of wire around the input lead of your receiver. Adjust the frequency-calibrate capacitor until a zero beat is obtained between the crystal signal and the WWV signal. In so doing, you have set your crystal oscillator precisely on 100 kHz. As a result, every 100-kHz calibrate point you receive over the low- and high-frequency spectra will be a very accurate one.

6

Special IC Systems

The science of miniaturization has allowed elaborate electronic systems to be reduced to more efficient monolithic structures. Two examples of such structures are miniaturized receivers and transmitters which require a minimum of external discrete components. Another example is waveform generators capable of generating a variety of waveforms. Furthermore, information can be placed on these waveforms using a variety of modulation methods. Phase-locked-loop (PLL) systems are another case where one package includes the facility for setting up a variety of modulation, demodulation, and frequency-control systems.

Digital integrated circuits are a special form of on/off devices used widely in computer systems, calibrators, test instruments, and counters. However, these digital devices do have applications in linear systems too, and a basic knowledge of their functions is important.

TRANSMITTERS AND RECEIVERS

The Lithic Systems LP2000 is an example of a microtransmitter requiring only a limited number of external components to complete a 50-mW output transmitter (Fig. 6-1). The radio-frequency section consists of an oscillator, two buffer stages, and the modulated-output amplifier. This section can be operated as a cw transmitter with a 100-mW power capability. Also included in the integrated circuit is a facility for power distribution and control, a modulation preamplifier and a modulator. Current demand from a 12-volt supply source is 50 mA for cw operation and 28 mA for a-m transmission.

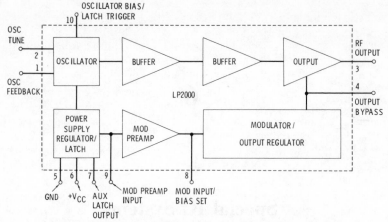

Fig. 6-1. Microtransmitter made by Lithic Systems, Inc.

External components required are microphone and resistor networks for controlling modulation level and idling current. Also needed are oscillator and output tuned circuits plus the antenna-matching facility.

A complete transmitter diagram is shown in Fig .6-2. The crystal oscillator and its tuned circuit are connected to terminals 1, 2, and 10. Turn-on bias is applied by way of terminal 10. This single-pole double-throw switch is used to turn the complete transmitter on and off.

The output-tuning and -matching network is connected to terminal 3, the collector circuit of the output transistor. A capacitive divider provides impedance matching to a low-impedance antenna system.

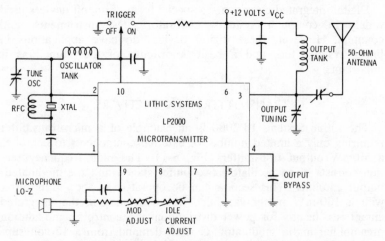

Fig. 6-2. Transmitter circuit conguration using an LP2000 microtransmitter.

The microphone signal is applied to the base of the input transistor of the modulation preamplifier by way of terminal 9. A resistive divider network connected to terminals 7, 8, and 9 controls the modulation level and the no-modulation idling current.

A schematic of the internal components is given in Fig. 6-3. Transistor Q8 at the very center of the diagram is the oscillator. Frequency-determining elements are connected externally between its terminals 1 and 2. The turn-on bias is applied to its base by way of terminal 10. An unusual aspect of the oscillator transistor is that it draws a fixed current determined by the temperature-compensated base-bias network and the constant-current source (transistor Q7) connected to its emitter. As a result, parasitic capacitance, gain, and junction temperature are held constant. Oscillator stability is strictly a function of the external frequency-determining components. The oscillator operates on both fundamental and overtone crystals.

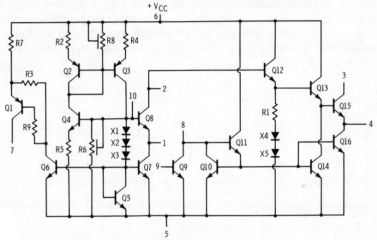

Fig. 6-3. Internal schematic of the LP2000.

The buffer stages consist of two emitter-followers, transistors Q12 and Q13. The emitter load of transistor Q12 is low-value resistor R1 and diodes X4 and X5. The second buffer has a transistor for an emitter load. For cw, its base bias is held constant. However, with a-m modulation, it is made to vary, introducing a modulation component into the emitter circuit. As is customary in transistor modulation systems, it is usually advisable to modulate the driver (last buffer) and the output stage.

The output stage, transistor Q15, acts as a grounded-emitter circuit insofar as the radio-frequency signal is applied to its base. However, the modulating wave is applied to the base of its constant-current source, transistor Q16.

The modulating wave is applied to the base of transistor Q9 by way of terminal 9. The wave is then processed by the pair of feedback transistors, Q10 and Q11. From there it goes onto the bases of the modulator transistors Q14 and Q16.

Modulation level and idling current adjustment are established when the resistive path is closed between terminals 7, 8, and 9. Transistor Q1 serves as a regulated collector load for transistors Q9 and Q10. The resistor connected between terminals 7 and 8 (Fig. 6-2) is able to control the biasing of transistors Q14 and Q16. Inasmuch as they are the bias sources for buffer and output transistors Q13 and Q15, the idling current can be preset. The adjustable resistance between terminals 8 and 9 determines the biasing of the input transistor Q9 and, therefore, gain and modulating level can be adjusted.

A most unusual integrated circuit is the simple three-terminal device by Ferranti Semiconductors, shown in Fig. 6-4. You apply a radio-frequency signal to the input (frequency range between 200 kHz and 1.5 MHz), and you obtain enough demodulated audio output to drive a headset or a simple follow-up audio amplifier. Output is typically 30 mV.

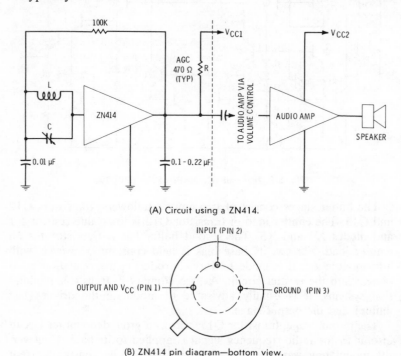

(A) Circuit using a ZN414.

(B) ZN414 pin diagram—bottom view.

Fig. 6-4. Ferranti Semiconductors IC a-m radio receiver.

Internally, the device consists of a 10-transistor tuned radio-frequency circuit for a-m detection and demodulation. Current consumption is 1 mA. The device operates with a supply voltage of 1.1 to 1.8 volts. A simple resistive divider can be set up when the circuit is to be used with a 6- or 9-volt battery. Details are given in Chapter 7.

External components required are the tuned resonant circuit, two fixed capacitors, two resistors and a headset or audio amplifier. The device includes a built-in agc system. Total harmonic distortion at the output does not exceed 2 percent. The circuit will demodulate a 100-μV signal across the input of the antenna coil. If the input resonant circuit has a Q of 100 or more, the circuit will respond to a 10-μV input signal.

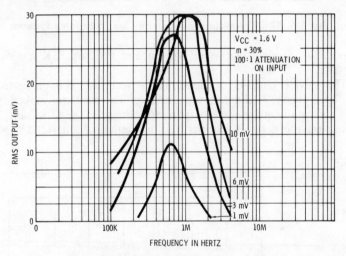

Fig. 6-5. ZN414 frequency-response curves.

Bandwidth measurements of the device are shown in Fig. 6-5. Note that a 10-μV input signal in particular is able to produce a usable output as high as 5 MHz. The performance peaks over the a-m broadcast band spectrum.

WAVEFORM GENERATORS

Waveform generators of various types have now been included in integrated circuits. A unique one is the XR-205 made available by Exar Integrated Systems. It is a 16-pin device (Fig. 6-6) that generates sinusoidal, triangular, square, sawtooth, ramp, and pulse waveforms. Furthermore, these various waveforms can be amplitude modulated, frequency modulated, or modulated with a combination of a-m and fm. Double sideband, as well as suppressed-carrier a-m,

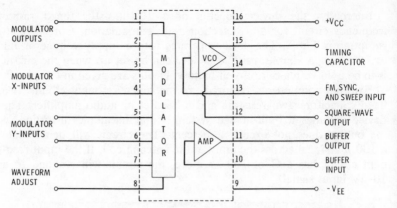

Fig. 6-6. XR-205 waveform generator by Exar Integrated Systems.

is possible. Straight cw on/off keying and FSK or PSK modes are possible.

The device consists of three major sections: voltage-controlled oscillator (vco), modulator, and amplifier. The oscillator makes available sine- and square-wave outputs. Frequency range of its sine- and square-wave generation extends from a fraction of a hertz to 5 MHz.

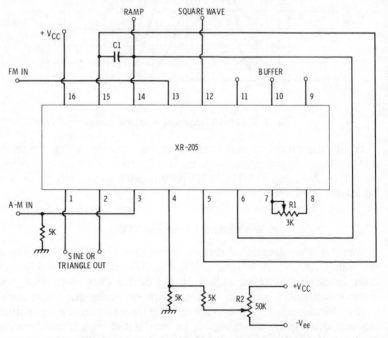

Fig. 6-7. Connecting the XR-205 as a function generator.

Triangle and ramp waveforms are feasible up to a maximum of about 500 kHz.

As shown in Fig. 6-7, the frequency of the oscillator is controlled by applying a suitable dc voltage to pin 13. This can be a modulating wave as well and will produce frequency modulation of the oscillator. The oscillator frequency is set by capacitor C1, connected between terminals 14 and 15. Output is taken from between these two terminals and is applied to the modulator input by way of terminals 5 and 6. Ramp and square waves can be removed at terminals 12 and 14.

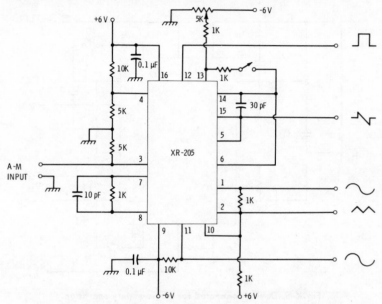

Fig. 6-8. The XR-205 connected for high-frequency operation.

The output of the modulator can be either sinusoidal or triangular as determined by the setting of the potentiometer connected between terminals 7 and 8. This adjustment does the wave shaping. An amplitude-modulated output is obtained by applying the modulating wave to pin 3. The level of the output signal is controlled by setting the bias at terminal 4.

To obtain isolation or a higher output, a buffer amplifier can be connected. This device is included internally. Connection can be made by joining terminals 1 and 2 to terminals 9 and 10. Output is then removed at terminal 11. The output voltage swing is several volts peak-to-peak.

The circuit of Fig. 6-8 has been designed for higher frequency operation up to at least 10 MHz. Output at this frequency is approximately 700 mV. Note the small value capacitor between terminals 14 and 15. Single power-supply operation, up to a frequency limit of 3-4 MHz with an output voltage of 2-3 volts, can be obtained with the circuit of Fig. 6-9.

Two such devices can be combined into a single generator, including both radio-frequency and modulating-wave source in a single module (Fig. 6-10). The device at the left of the circuit provides the modulating signal (sine, square, triangle, or ramp) as selected by switch S_A. Capacitor C1 determines the operating frequency. A

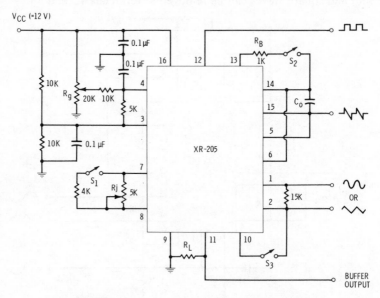

Fig. 6-9. Device connected for single-supply operation.

variable capacitor will permit a means of frequency adjustment. Switch S_B sets the modulating-wave source for either a 20- or 50-percent duty cycle. Switch S_C determines whether the carrier generator will be unmodulated, frequency modulated, or amplitude modulated. Potentiometer P1 sets the level of the modulating wave; P2 is a zero-carrier adjustment for suppressed-carrier modulation, and potentiometer P3 sets the shape of the modulating wave.

The second device operates as a carrier generator. It also includes switch S_D for setting the duty cycle. Switch S_E selects the type of output waveform, while capacitor C2 determines the carrier frequency. Potentiometer P5 is for wave shaping.

This versatile waveform generator can be used as a signal source for a transmitter, or as a very effective piece of transmitter and transceiver test equipment. Such a device will be constructed as a project in Chapter 10.

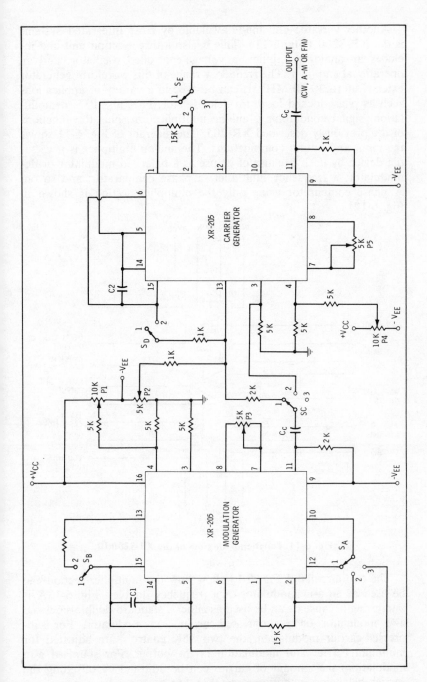

Fig. 6-10. An am/fm generator.

Another versatile unit made available by Exar Integrated Systems is the XR-S200 (Fig. 6-11). This is also a three-section unit and includes an analog multiplier, a voltage-controlled oscillator, and an operational amplifier. The frequency range of this waveform generator extends up to 30-40 MHz. It can be used in a variety of applications such as phase-locked loop, fm demodulator, FSK and PSK demodulation, signal conditioning, analog multiplication, plus the functions of the previously described XR-205. The diagram of Fig. 6-11 shows the input and output combinations. The analog multiplier is a versatile device by itself because of its use as a balanced modulator or demodulator, a frequency multiplier, a phase comparator, and so on. A phase comparator using only the multiplier section is shown in Fig. 6-12.

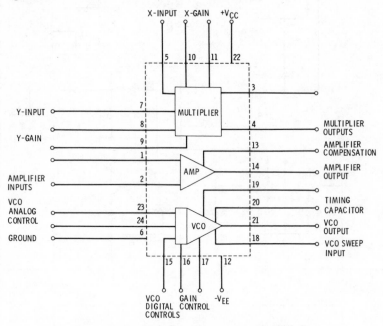

Fig. 6-11. Functional diagram of the XR-S200 IC.

The two circuits of Fig. 6-13 show how the multiplier section can be used as an a-m modulator or a frequency doubler. Fig. 6-13A is interesting because it can be used as either a standard double-sideband a-m modulator or a suppressed-carrier a-m modulator. For suppressed-carrier modulation, the two 15K controls are adjusted for minimum carrier and modulation in the output. For standard a-m modulation, the amount of carrier can be adjusted by offsetting the carrier balancing potentiometer (R2).

The source of carrier for these operations can be the voltage-controlled oscillator. Inasmuch as the oscillator can be frequency modulated, it is no problem to generate an fm signal too. If desired, the vco can be crystal controlled. A complete a-m/fm signal source is shown in Fig. 6-14. (The output of the crystal-controlled oscillator can be either amplitude or frequency modulated.)

One can expect a definite trend toward the use of waveform generators for many applications. When used as a waveform generator, the XR-S200 forms the same variety of signals as the XR-205.

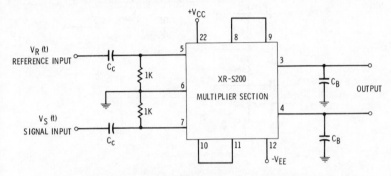

Fig. 6-12. A phase comparator using only the multiplier section of an XR-S200 IC.

PHASE-LOCKED LOOP (PLL)

Many applications for phase-locked loops include frequency multiplication and division, demodulation without tuned circuits, signal conditioning and interference suppression, frequency synthesizers, data synchronizers, tracking filters, decoders, modulators and demodulators, and so forth.

The basic phase-locked loop (Fig. 6-15) has three main sections—phase comparator, low-pass filter, and voltage-controlled oscillator—all connected in a closed-feedback loop. Almost an entire PLL circuit can be included in an integrated circuit. Its major objective and big advantage in many applications is its ability to lock onto the frequency of any applied signal.

To understand the operation of the PLL, let us assume first that the device is not locked onto the incoming signal, even though the two frequencies are reasonably close in frequency. The incoming signal and the locally generated signal from the vco (e_i and e_{osc}) are compared in the phase detector. The output of the phase detector will be a difference or beat frequency (e_x). Since it is low in frequency, output e_x is able to pass through the filter and pass to the voltage-controlled oscillator. The passband of the filter is such that the incoming and vco components do not pass. The feedback component e_f then shifts the frequency of the oscillator and locks it onto the incoming frequency.

Actually, it is a dc voltage that holds the oscillator on frequency because at the moment lock-in occurs, both the input signal and the vco signal are on the same frequency; thus, the output of the phase

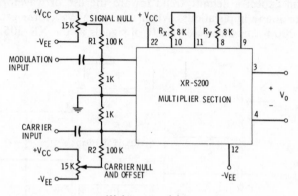

(A) An a-m modulator.

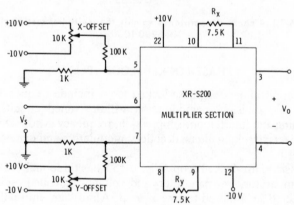

(B) A frequency doubler.

Fig. 6-13. Two additional uses for the XR-S200.

detector is zero. If either the incoming signal changes frequency or if the vco attempts to drift in frequency, there is an instantaneous shift in the phase between the two components in the phase detector. Immediately, there will be a dc output from the phase detector which will be either positive or negative, depending on the direction of the frequency separation. This dc output will cause a frequency-control voltage to be immediately applied to the voltage-controlled oscillator (vco) that will again equalize the two frequencies.

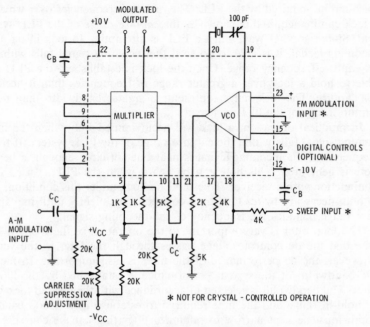

Fig. 6-14. An am/fm signal source.

If it is the vco that attempts to drift in frequency, the dc voltage will prevent it from doing so. If it is the incoming signal that changed in frequency, the dc voltage will cause the vco to follow that change in frequency. From this it is apparent that the phase-locked loop can be made to hold its own frequency precisely or be made to follow the frequency of an applied signal. This action has many attractive attributes in various types of electronic systems. A number of these will be described.

The range of frequencies over which a PLL can hold a lock is known as the *lock range* or *lock-in range* of the system. The lock-in range is always greater than the range of frequencies over which the

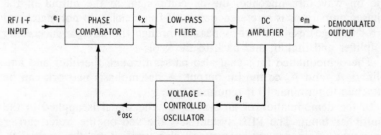

Fig. 6-15. Block diagram of a PLL feedback loop.

lock can be acquired by the PLL. The range of frequencies over which a lock can be acquired is known as the *capture range* of the PLL system. Stated another way, when a PLL is attempting to lock in on an incoming signal, it is able to do so if the incoming signal falls within its capture-frequency range. Once the lock is established, the PLL is able to hold a lock over a greater range of frequencies than it needs for capture. The lock range represents a greater bandwidth than the capture range of the PLL system.

In practical operation, a lock will occur immediately when the incoming signal falls within the capture range of the PLL system. If the frequency of the incoming signal is near but not close enough, a beat note is generated. This note can be heard when the PLL is used in conjunction with a receiver. To establish a lock under this condition, it is only necessary to readjust the vco frequency slightly until its capture range includes the frequency of the incoming signal.

The loop filter is very important to the operation of the PLL. It is true that the dc control voltage passes through the filter to the vco. However, the ac performance of the filter is also important. In fact, the bandwidth of the system performance is established by the loop filter. The bandwidth is selected in accordance with the desired lock-in or hold-in range and the time required to establish lock. Loop bandwidth must be kept narrow to minimize jitter that can be caused by external noise or interference components. However, the loop bandwidth must be adequate to pass desired components and establish an adequate capture range. In all circumstances, the loop bandwidth must be wide enough to accommodate the frequency components that make up the modulation. When used for the demodulation of an fm signal, one understands that the bandwidth should be greater than that required for the demodulation of a narrow-band a-m signal.

A function block diagram of the Signetics Corporation NE561B PLL integrated circuit is shown in Fig. 6-16. This device is an am/fm demodulator system. The vco is an emitter-coupled multivibrator, the frequency of which can be controlled by an external capacitor C_O. Its output is applied to the phase comparator along with the incoming fm radio-frequency signal. When the vco and the center frequency of the fm wave are matched, the dc component at the output of the phase comparator is a reference zero. If frequencies do not match, an error voltage results which is passed through the filter, a succeeding amplifier and limiter, and back to the vco.

The demodulated fm signal also passes through the filter and amplifiers A_1 and A_2 to the fm output. A de-emphasis network can be attached to terminal 10 if required.

In the demodulation of an a-m sgnal, the signal is applied to the multiplier input. The PLL system sets the vco on the exact carrier frequency. This vco component is also applied to the multiplier.

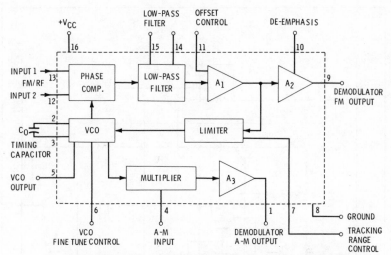

Fig. 6-16. Block diagram of the Signetics Corporation NE561B phase-locked loop.

Therefore, the multiplier acts as a product detector, and its output frequency is the difference between the a-m input signal and the locally generated carrier from the vco. The demodulated a-m signal is increased in level by amplifier A_3 and is recovered at terminal 1 of the device.

A simplified schematic of the internal make-up of the NE561B PLL is given in Fig. 6-17. The phase comparator shown on the right side of the diagram is a doubly balanced modulator. The two upper pairs of transistors (Q17, Q18 and Q19, Q20) function as a synchronous switch driven by a signal from the voltage-controlled oscillator. The fm signal is applied to the bottom pair of the transistors (Q21 and Q22). The fm signal varies the relative emitter currents of the balanced modulator, producing a difference frequency at the output. The phase comparator circuit is stabilized by the constant-current transistor Q23.

The differential voltage variation at the output of the phase comparator is a copy of the frequency modulation on the incoming signal. The demodulated audio is removed from the phase comparator section by way of emitter-follower transistor Q16. A common-emitter amplifier builds up the signal level and applies it to the emitter-follower output stage, transistor Q25. The de-emphasis network, if needed, can be connected between the two stages. The emitter-follower output stage provides a low-impedance output for driving any succeeding audio module.

The multiplier section is shown at the top left of the diagram and is also a doubly balanced modulator with the a-m signal applied dif-

203

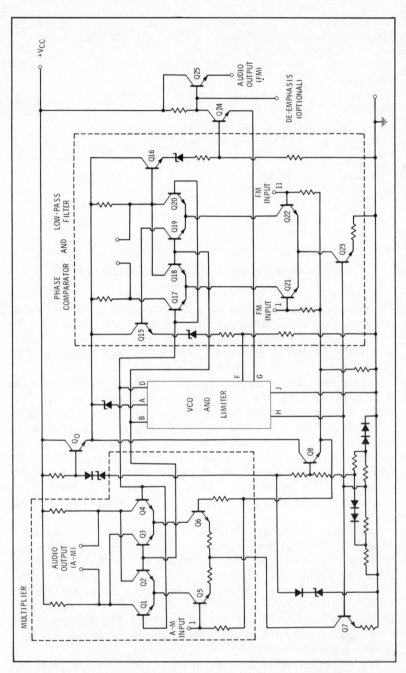

Fig. 6-17. Simplified schematic of the internal circuit of the NE561B.

ferentially to transistors Q5 and Q6. The local-oscillator signal from the voltage-controlled oscillator is applied differentially to the two differential pairs. Again the incoming audio signal varies the relative emitter currents, producing a differential output corresponding to the audio modulation on the incoming signal. It is to be noted that a dc control voltage is transferred to the phase comparator, keeping the frequency of the voltage-controlled oscillator identical to that of the incoming a-m carrier.

A simplified schematic of the voltage-controlled oscillator (Fig. 6-18) shows its general arrangement and the influence of its timing

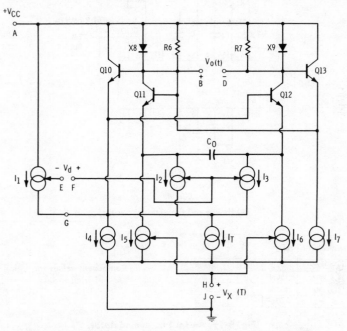

Fig. 6-18. Internal circuit of the vco section of the NE561B.

capacitor (C_O) on operating frequency. It is an emitter-coupled multi-vibrator with the output taken off differentially between terminals B and D. The frequency-controlling elements are in the emitter circuit and consist of the external timing capacitor and constant-current sources I_2, I_3, I_5, and I_6. The frequency of the multivibrator is also controlled by a voltage placed between terminals E and F. When there is no control voltage, the circuit is balanced with the frequency set by the previously mentioned frequency-determining capacitor and currents. The presence of voltage causes a differential current partitioning between I_2—I_3 and I.

The limiting action of the PLL is a built-in feature of the vco and is set by the current divisions between the control currents I_2, I_3 and the bias sources I_5, I_6. The tracking range of control can be set externally between terminals H and J, which influences the partitioning of the currents I_2 and I_5.

A complete external circuit for using the device as an fm demodulator is shown in Fig. 6-19A. The fm signal is applied between termi-

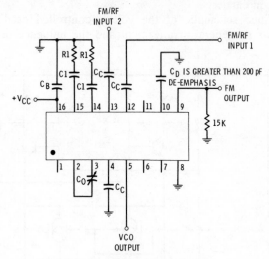

(A) Device and the external circuitry.

FREQUENCY	C_O	R1	C1
10.7 MHz	30 pF	100 Ω	1000 pF
4.5 MHz	54 pF	50 Ω	2000 pF
455 kHz	650 pF	0	5000 pF
67 kHz	4600 pF	0	0.01 μF

(B) Component values for typical i-f frequencies.

Fig. 6-19. An fm PLL demodulator.

nals 12 and 13. Capacitor C_O determines the vco frequency. An approximate equation for determining its value is:

$$C_O = \frac{300}{f_o}$$

where,

C_O is the capacitance, in microfarads,
f_o is the oscillator frequency, in hertz.

Values for typical i-f frequencies are given in the chart of Fig. 6-19B. If the vco is to be used as a tunable local oscillator, this capacitor is made variable.

The values of the series resistor-capacitor combinations connected to terminals 14 and 15 determine the characteristics of the low-pass loop filter. The time constant of the filter is:

$$tc = C1(R1 + R_{in})$$

where,

R_{in} is the internal impedance and is approximately 6000 ohms.

A practical version of the equation would be:

$$C1 = \frac{1}{2\pi f_h R1}$$

where,

f_h is the desired bandwidth of the filter.

For example, frequency f_h would be 15 kHz for use in an fm receiver.

A capacitor can be used to ground one side of the input to permit the fm signal to be applied single-ended and differential.

Capacitor C_D provides de-emphasis, and its value is based on the time constant:

$$tc = R_D C_D$$

Resistor $R_D = 8000$ ohms, the output impedance at the de-emphasis terminal. The time constant used is either 75 microseconds for fm broadcasting, or whatever value is needed for a particular fm demodulation service.

The tracking range of the device, based on the internal adjustment, is approximately 35 percent of the vco frequency. The vco will follow over this frequency range once a lock-in has been established. It can be changed upward or downward by controlling current at pin 7. The chart of Fig. 6-20A shows how the current at pin 7 influences

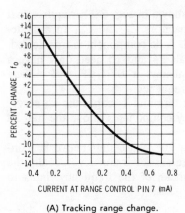

(A) Tracking range change.　　(B) Change of vco frequency.

Fig. 6-20. Characteristic current curves.

the percentage change in the frequency. According to the chart, the required current is 0.15 mA.

Fig. 6-20B shows the influence of the fine-tuning current at pin 6 on the oscillator frequency. The specified current value changes the frequency of the vco according to the percentage figure given on the chart. An introduced current of 5 mA produces a 25-percent change in frequency.

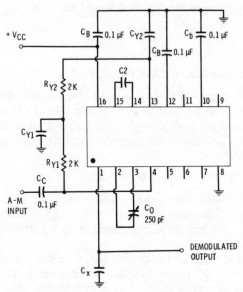

Fig. 6-21. An a-m PLL demodulator.

The phase-lock a-m demodulator, shown in Fig. 6-21, requires that the fm inputs be ac grounded by capacitors C_B and C_Y. In a-m demodulation, a pass filter is not needed. A dc component is transferred. Capacitor C2 across the filter terminals 14 and 15 need only have a low reactance at the operating frequency; a $0.01-\mu F$ will suffice. Capacitor C_O again determines vco frequency.

Demodulated output is removed at terminal 1; capacitor C_x serves as the output filter removing frequency components above the highest modulating frequency.

Capacitors C_Y and R_Y operate as a 90° phase-shift network. Constants must be chosen according to the carrier frequency. Usually the frequency selected is the geometric mean of the frequency limits to be tuned, or:

$$f_c = \sqrt{f_h f_l}$$

In the case of the a-m broadcast band, this mean frequency would be approximately 0.94 MHz. Again the time constant would be:

$$tc = C_Y(R_Y + R_0)$$

Again, R_0 is the output resistance and equals 8000 ohms.

In a practical case, assuming an R_Y value of 3000 ohms, the required value for the capacitor C_Y becomes:

$$C_Y = \frac{1.3 \times 10^{-4}}{f_c}$$

where,
C_Y is the capacitance in pF,
f_c is the carrier frequency in MHz.

Some of the waveform-generator ICs can also be used as phase-locked loops. An example is the XR-S200 of Fig. 6-22. Here, it is being used as a PLL fm detector. In this arrangement, the multiplier section is connected as a phase detector and its output is applied to the voltage-controlled oscillator by joining terminals 4 to 23 and 3 to 24. The low-pass filter connects to the same outputs. Demodulated

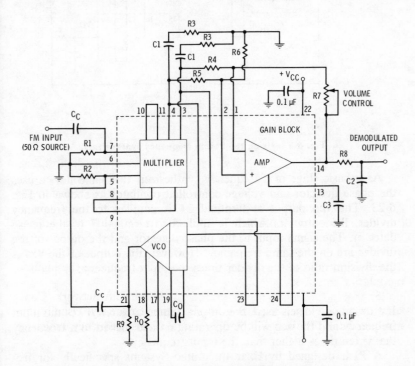

Fig. 6-22. PLL fm detector using an XR-S200 integrated circuit.

209

audio signal from the same point is applied in differential to the operational amplifier gain block. Output is removed at terminal 14.

The volume control is in the feedback link of the operational amplifier and connects back to the inverting terminal of the operational amplifier. Capacitor C3 provides compensation for the operational amplifier, while the R8-C2 combination provides 75-microsecond deemphasis. The vco component for the phase detector is coupled from terminal 21 through a capacitor to terminal 5 of the phase detector.

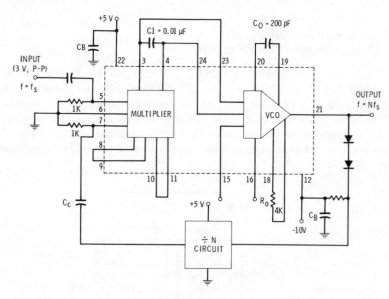

Fig. 6-23. Using the PLL in frequency synthesizers.

As a basic stage of a frequency synthesizer, this same device uses the phase detector and voltage-controlled oscillator as shown in Fig. 6-23. The vco output at terminal 21 is coupled to the frequency divider, the output of which is applied to terminal 7 of the phase detector. The signal input to the phase detector and the output of the divider are on the same frequency. However, the output of the vco is the division ratio of the divider times the input frequency:

$$f_o = Nf_s$$

If a decade divider is used, the output frequency is ten times the input frequency and the vco will be operating on and locked to a frequency that is ten times higher than the input frequency.

A PLL designed by Exar Integrated Systems specifically for frequency-shift keying (FSK) detection uses a vco, an operational ampli-

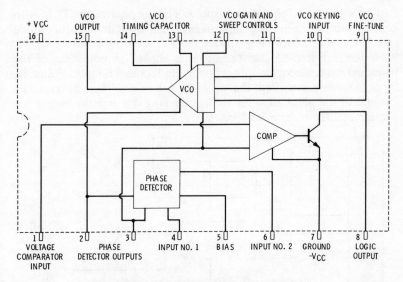

Fig. 6-24. A PLL FSK modulator/demodulator.

fier, plus a comparator and a phase detector (Fig. 6-24). In an FSK demodulation system, there are two incoming frequencies, one repre-

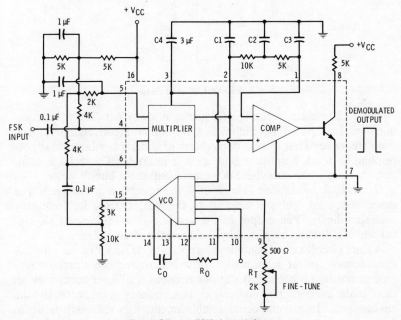

Fig. 6-25. An FSK demodulator.

211

senting a mark and the other a space. The purpose of the PLL is to convert this two-frequency input signal to a demodulated amplitude signal of two specific voltage levels.

Notice in Fig. 6-25 that the signal is applied to one input of the phase detector; the vco output is applied to terminal 6 input. When the input frequency is shifted, the polarity of the dc voltages at terminals 2 and 3 of the phase detector shift, producing one polarity output for a space and the reverse for a mark.

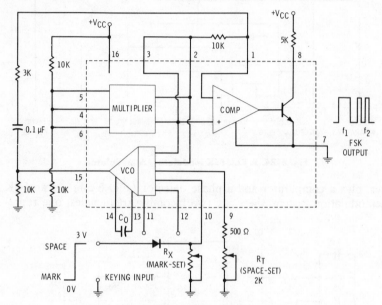

Fig. 6-26. An FSK modulator.

The PLL maintains the vco frequency midway between the incoming mark and space frequencies. In this manner, a conversion is made from frequency shift to two dc voltages of different polarity that correspond to mark amplitude and space amplitude. The voltage comparator simply converts these dc voltage shifts to a binary pulse.

In the FSK modulator (Fig. 6-26), the operation is reversed. The space and mark pulse is applied to the vco, shifting its frequency correspondingly. The output of the vco is a two-frequency FSK signal which is further amplified by the operational amplifier.

A final circuit uses a Signetics Corporation NE565 PLL as an SCA demodulator. In the transmission of uninterrupted commercial background music, fm broadcast stations transmit a 67-kHz subcarrier on their main carrier. This subcarrier is frequency modulated by the music signal. In a commercial establishment, it is removed by a decoder that extracts the 67-kHz component from the composite fm broadcast signal. The PLL circuit of Fig. 6-27 requires no tuned circuit in this application. The vco is set to 67 kHz and is locked in by the incoming 67-kHz component. The fm sidebands, however, are demodulated and, through a suitable filter, applied to the audio input of a commercial sound system. The 5K potentiometer permits the vco to be locked exactly on frequency. The frequency response of the circuit shown extends up to 7000 hertz.

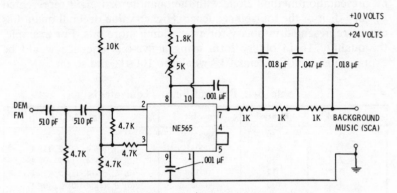

Fig. 6-27. An SCA demodulator using an NE565.

DIGITAL INTEGRATED CIRCUITS

Digital ICs are found in many linear systems. They are often a part of the associated accessories and test equipment. Some of these applications are calibrators and counters, frequency synthesizers, indicating meters and instruments, multipliers and dividers, scanners, keyers, and remote control and switching systems. A first step in understanding digital ICs is to learn just a bit about the binary numbering system.

Binary Numbering

In a binary system, you can count with only two numerals, zero and one. In both the binary and decimal languages, nothing is zero or 0. Likewise, the quantity one is the same for both languages and is written as one or 1. From this point on, things differ. In decimal language, the quantity two is written as 2; in binary language, it is written as 10. Very often, the binary code used is in the form of a four-bit series, and the quantities zero, one, and two are set down as follows:

$$zero = 0000$$
$$one = 0001$$
$$two = 0010$$

The four-bit expression for zero is telling us that there is *no* eight, *no* four, *no* two and *no* one. For the quantity one, it is saying that there is *no* eight, *no* four, *no* two and *one* one. Likewise, for the quantity two, the expression says that there is *no* eight, *no* four, *one* two and *no* one.

Thus, in four-bit binary language, the quantity nine is written as 1001, telling that there is *one* eight, *no* four, *no* two and *one* one. One 8 plus one 1 equals the quantity 9. Table 6-1 makes a comparison between the decimal and binary presentations. It also shows the four-bit presentation method along with the quantity or *weight* represented by each digit of the four-bit sequence. Higher-value decimal quantities can be represented in binary form by adding more bits. For example, the quantity 16 in binary form using a five-bit sequence would be expressed as 10000; decimal 17 would be 10001; and so on.

Table 6-1. Decimal-Binary Equivalents

Decimal	Binary	Four-bit	Weight 2,4,2,1
0	0	0000	(0) 8, (0) 4, (0) 2, (0) 1
1	1	0001	(0) 8, (0) 4, (0) 2, (1) 1
2	10	0010	(0) 8, (0) 4, (1) 2, (0) 1
3	11	0011	(0) 8, (0) 4, (1) 2, (1) 1
4	100	0100	(0) 8, (1) 4, (0) 2, (0) 1
5	101	0101	(0) 8, (1) 4, (0) 2, (1) 1
6	110	0110	(0) 8, (1) 4, (1) 2, (0) 1
7	111	0111	(0) 8, (1) 4, (1) 2, (1) 1
8	1000	1000	(1) 8, (0) 4, (0) 2, (0) 1
9	1001	1001	(1) 8, (0) 4, (0) 2, (1) 1
10	1010	1010	(1) 8, (0) 4, (1) 2, (0) 1
11	1011	1011	(1) 8, (0) 4, (1) 2, (1) 1
12	1100	1100	(1) 8, (1) 4, (0) 2, (0) 1
13	1101	1101	(1) 8, (1) 4, (0) 2, (1) 1
14	1110	1110	(1) 8, (1) 4, (1) 2, (0) 1
15	1111	1111	(1) 8, (1) 4, (1) 2, (1) 1

Various codes based on the binary 1 and 0 concept have evolved to meet the requirements of digital equipment operation and objectives. A simple code that has been used extensively is known as the binary-coded-decimal (BCD) method. Four binary bits are employed. Each is said to have a weight of 8, 4, 2, 1 in the order of digits from left to right. Each digit position has a definite value (weight).

In the true binary case, it is simply 8, 4, 2, 1 as shown in Table 6-1. Conversion to decimal values involves the simple addition of the weights of the digits as shown previously. For special needs, there are various other types of codes, some involving more than four digits or characters. In the case of BCD, however, a simple four-bit number is used to express all decimal quantities zero through nine. Although a four-bit number can designate higher numbers (up to 15 as shown in the chart), the BCD code restricts each four-bit character to decimal numbers from zero through nine.

When a higher number is to be indicated in binary form using the BCD code, additional four-bit characters are conveyed. For example, the number 35 in the BCD code becomes 0011   0101.

$$3 \quad 5$$
$$0011 \quad 0101$$

Note that the first four-bit character is the binary representation of three while the second is the representation of the decimal quantity five. Thus, the number 6751 would be written:

$$0110 \quad 0111 \quad 0101 \quad 0001$$

The intriguing part of the binary presentation is that numbers can be represented by combinations of binary 0 and binary 1. Furthermore, it is very easy to understand how to use a simple switch to see a binary 1 or a binary 0. For example, a closed switch could represent a binary 1; the same switch when open, a binary 0. A sequence of switches could then be used to establish a complete binary counting system (Fig. 6-28). The closed position of the switch in the cir-

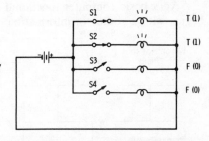

Fig. 6-28. Binary number representation using switches and lamps to show the decimal number 12.

cuit is customarily called the *on* position; the open position is the *off* position. In computer vernacular, the closed position could designated *true* or binary logical 1, while the off position of the switch would be designated *false* or logical 0. In representing the decimal quantity twelve, switches 1 and 2 would be closed, while switches 3 and 4 would be open, setting up the sequence of 1100.

Logic Circuits

In an actual computer, the switches would be electronic and not mechanical. They operate at very high speeds causing many operations to occur simultaneously and at a high repetition rate.

Some considerable confusion exists because of the several ways of designating a logical 1 and a logical 0. In general, the terms yes and no, true and false, as well as 1 and 0, are identical. In actual circuits, logical 1's and 0's are represented by voltage levels. In a negative logic circuit, the most positive voltage level (high) is 0 and the most nega-

215

tive level (low) is 1. Conversely, in a positive logic circuit, the most positive level (high) is defined as a logical 1, while the most negative voltage level (low) is a logical 0.

It follows, then, that in a positive logic circuit, the following terms are one and the same:

1 and 0

yes and no

true and false

high and low

Conversely, for negative logic circuits the identities are:

1 and 0

no and yes

false and true

low and high

The most common method used is positive logic.

A very basic computer functional diagram is given in Fig. 6-29. It begins with a source of information or *data input* that is to be calcu-

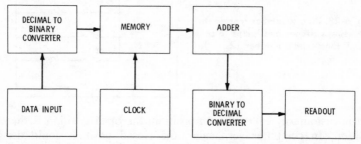

Fig. 6-29. Block diagram of the functional operation of a computer.

lated or measured. This could be the keyboard of one of the small handheld electronic calculators. Inside of the device is a *converter* that changes the decimal quantities that are tapped out on the keyboard to binary quantities. These are applied to a *memory section* which stores all the information being fed in. At an appropriate time, all of this stored information is applied to the *calculating section* of the computer. Under control of a *timing* or *clock signal,* the actual calculation is performed; be it addition, subtraction, multiplication, division, etc. All of this is accomplished with on/off electronic switching and the complex juggling of binary 1 and binary 0 numbers.

The binary answer is supplied to an *output converter* which changes the binary quantity back to a decimal quantity. Converter output is supplied to a *display readout* device which furnishes the answer in the form of a decimal quantity. The display device is usually a segmented electronic glow device that lights up in group fashion in accordance with the voltage level presented to it by the converter. The voltage level is, in turn, a function of the particular decimal number the converter is supplying at a particular instant.

In a very abbreviated and simple explanation, a computer goes through the following operations—insertion of decimal information, conversion of decimal form to binary form, storage of information, calculation of binary answer, conversion of binary answer to decimal form, and display of decimal answer. The secret of a computer's ability to perform the high-speed and complex operations is the ability of the system to employ an elaborate system of exceedingly fast on/off switching. The circuits that perform the above operations can be included in digital integrated circuits. This results in a great saving in space, and it permits speeds of operation that now have a switching rate capability of more than 100 million bits per second. Consider now some of the basic digital circuits.

BASIC LOGIC BLOCKS

Five of the basic logic functions are referred to as AND, OR, NOT (inverter), NOR, and NAND circuits. These logic functions are depicted in Figs. 6-30 through 6-34. Each illustration includes a simple "switch and lamp" schematic, the block representation of the logic function, and a truth table. Several actual digital integrated circuits are described later to show how these functions are accomplished electronically.

An OR gate has one output but two or more inputs. When the battery voltage (a logical 1 assuming a positive logic) is applied to either or both inputs of Fig. 6-30A by closing appropriate switches, the output voltage is also at a logical 1 level. The lamp glows, indicating a true condition. If no voltage is applied to either input by keeping both switches open (logical 0 voltage), the output is also a logical 0. The lamp does not light, indicating the false condition.

The symbol for an OR gate circuit is shown in Fig. 6-30B. Two truth tables for an OR gate are given in Fig. 6-30C. The one on the left displays the circuit function in terms of *true* and *false* conditions. The second is the universal truth table for an OR gate circuit (using 0s and 1s). Some truth tables will use *yes* and *no* or *high* and *low*. As mentioned previously, all have the same meaning. Note that for the *true* input condition with either or both switches closed, the lamp glows, producing the *true* output. Conversely, when both switches are held open, the condition is false and the output is also false; the lamp remains off.

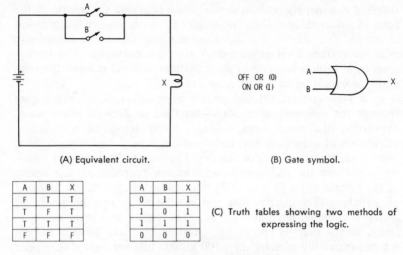

(A) Equivalent circuit.

(B) Gate symbol.

A	B	X
F	T	T
T	F	T
T	T	T
F	F	F

A	B	X
0	1	1
1	0	1
1	1	1
0	0	0

(C) Truth tables showing two methods of expressing the logic.

Fig. 6-30. Basic OR gate.

An AND-gate circuit also has two or more inputs and one output. Note that the switches are arranged in series in the circuit of Fig. 6-31A. In an AND-gate circuit, a logical 1 output is obtained only when a logical 1 voltage is applied to both inputs (both switches are closed). Under this condition, the lamp lights. If either or both switches are in the logical 0 condition (open), the output voltage is also a logical 0. As a consequence, the bulb does not light.

The gate symbol is shown in Fig. 6-31B and the truth table for an AND gate is given in Fig. 6-31C. Compare the OR-gate and AND-gate truth tables; notice, in particular, the following conditions:

1. For an AND-gate circuit, the output is in the logical 1 state when all inputs are a logical 1 level. The output is a logical 0 when any one or all of the inputs are a logical 0.

2. For an OR-gate circuit, the output is in the logical 1 state when any one or all of the inputs are in the logical 1 state. The output is a logical 0, when all inputs are a logical 0.

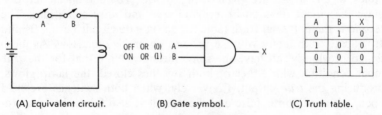

A	B	X
0	1	0
1	0	0
0	0	0
1	1	1

(A) Equivalent circuit.   (B) Gate symbol.   (C) Truth table.

Fig. 6-31. Basic AND gate.

A NOT circuit (Fig. 6-32) is usually called an *inverter* because the output is inverted relative to the input. The inverter has a single input and a single output. However, the output logic is the inversion of the input logic. If the input is true (a logical 1), the output is false (a logical 0). This condition is met when the switch is closed, or in its logical 1 state. If the input is a logical 0 (switch open), the output is a logical 1 and the lamp lights. Stated another way, if the input logic is true, the output is *not* true. If the input is false, the output is *not* false.

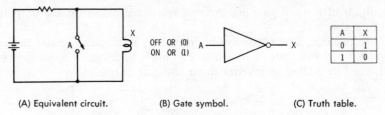

(A) Equivalent circuit.   (B) Gate symbol.   (C) Truth table.

Fig. 6-32. Basic inverter.

The gate symbol and truth table for an inverter are shown in Figs. 6-32B and 6-32C, respectively. Note that at the output of the inverter symbol there is a small circle. This is used to indicate a logic inversion by the NOT gate: 1 to 0 or 0 to 1.

A NOR gate is a combination of an inverter and an OR gate, while a NAND gate is a combination of an inverter and an AND gate. In a NOR gate (Fig. 6-33), a logical 1 state at either or both inputs (appropriate switches closed) does *not* produce a logical 1 state at the output. Rather the output is a logical 0 as shown in the truth table. Only when both inputs are a logical 0 is the output a logical 1. Note that the outputs shown in the truth table of a NOR gate are inverted with respect to the outputs shown in an OR-gate truth table. The NOR-gate symbol has a small circle at the output, indicating logic inversion.

When all inputs to a NAND-gate circuit are in the logical 1 state (switches closed), the output is *not* a logic 1. Instead, the output is a logical 0. In the circuit of Fig. 6-34A, the lamp does *not* light. When any one or all of the inputs are a logical 0, the output is in the logical 1 state, and the lamp in the circuit of Fig. 6-34A lights.

The gate symbol and truth table for a NAND gate are shown in Figs. 6-34B and 6-34C, respectively. Note again that the logic states for the

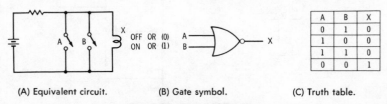

(A) Equivalent circuit.   (B) Gate symbol.   (C) Truth table.

Fig. 6-33. Basic NOR gate.

219

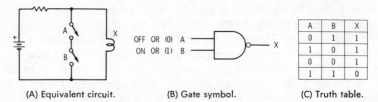

A	B	X
0	1	1
1	0	1
0	0	1
1	1	0

(A) Equivalent circuit.   (B) Gate symbol.   (C) Truth table.

Fig. 6-34. Basic NAND gate.

outputs of a NAND gate are inverted relative to those for the outputs of an AND gate.

In practice, a NOR gate can be formed by adding an inverter to the output of an OR gate as shown in Fig. 6-35. Likewise, a NAND gate is formed by adding an inverter to an AND-gate output.

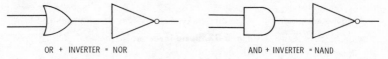

OR + INVERTER = NOR   AND + INVERTER = NAND

Fig. 6-35. Using an inverter to form NAND and NOR gates.

TTL CIRCUITS

A very popular form of digital integrated circuit is the 7400 series made by Texas Instruments Incorporated. An example of a NAND gate as used in this digital IC series is shown in Fig. 6-36. It has two inputs (A and B) to a multiemitter transistor (Q1). The collector output is direct coupled to the base of transistor Q2. In turn, transistor Q2 collector output is direct coupled to the base of transistor Q4 of the output drive system.

Normal supply voltage is 5 volts. Logical 1 output voltage is 3 volts; logical 0 output is 0.2 volt. Typical input and output pulses are shown

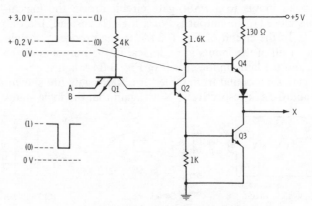

Fig. 6-36. A TTL 2-input positive NAND gate.

in Fig. 6-36. The input voltage logical 0 is slightly above the zero level results in a logical 1 output voltage of 3 volts. Conversely, a logical 1 input voltage results in a logical 0 output voltage of near 0.2 volt. Note that the output voltage is inverted in compliance with the logic of a NAND-gate circuit.

The multiemitter input transistor (Q1) is an npn type. A logical 0 (low) voltage applied to either one or both emitters causes the corresponding emitter-base junction of transistor Q1 to be forward biased. The collector current of Q1 is not adequate to forward bias transistor Q2. In fact, transistors Q2 and Q3 are held at cutoff. The collector voltage of transistor Q2 is positive with a logic 1 value, and is applied to the base of the output transistor Q4. Therefore, the output voltage of the circuit is 3 volts or high (a logical 1 state).

When both emitters of transistor Q1 are at the logical 1 state, the emitter-base junctions of Q1 are forward biased and cause a collector current to flow at a high enough level to forward bias and turn on transistors Q2 and Q3. The collector voltage of transistor Q2 drops as

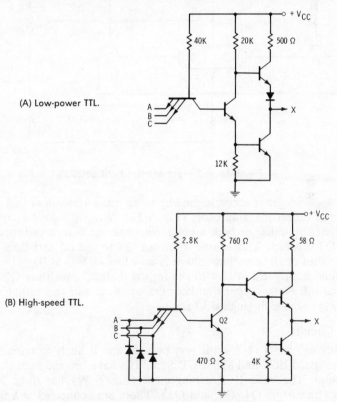

Fig. 6-37. Two 3-input positive NAND gates.

221

does the base voltage of transistor Q4. The output voltage drops to its logical 0 state of approximately 0.2 volt. The output is at a low impedance and can be *fanned out* to supply signals to a number of succeeding stages.

Low-power and high-speed versions of the same basic circuit are given in Fig. 6-37. In the circuit of Fig. 6-37A, the resistors are of a higher ohmic value; power dissipation is reduced. Power requirements may be as little as one-tenth of that drawn by the standard 7400 series circuits. Fig. 6-37B shows a high-speed version of the same circuit. Resistor values are in general lower. In addition, clamping diodes at the input discourage transients. Furthermore, the output combination is a Darlington pair which places a lighter load on transistor Q2 and minimizes the capacitance effects that can impede high-speed performance.

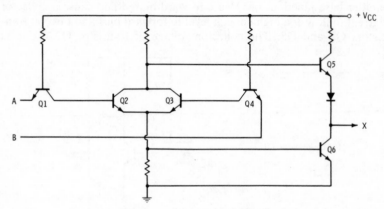

Fig. 6-38. A 2-input positive NOR gate.

NOR-gate logic is accomplished by using input transistors that feed a pair of differential transistors (Fig. 6-38). When a logical 1 voltage is applied to either or both input gates, either or both transistors Q2 and Q3 conduct. This causes transistor Q5 to cut off and transistor Q6 to turn on. The output voltage is now low (a logical 0).

When both gates are set to the logical 0 state, transistors Q2 and Q3 cut off. Now, output transistor Q5 conducts and the output voltage sets to the high (logical 1) state.

ECL Circuits

Although the TTL form is very popular and is likely to remain so, higher-speed ECL and MOSFET digital ICs have become increasingly common. The basic ECL arrangement (Fig. 6-39) has three input gates (transistors Q1, Q2, and Q3). These are connected as emitter followers. Therefore, the input impedance is high, and the circuit responds to fast input changes. Stability is enhanced by a constant-current source in the form of a high-value resistor (R_E) along with the reference transistor Q4. The emitter coupling also prevents the transistors from reaching saturation, making possible very high switching speeds.

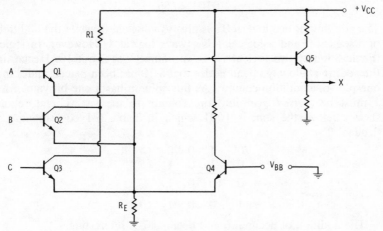

Fig. 6-39. Schematic for a basic ECL circuit.

When all the inputs operate in the logical 0 state (near cutoff potential), no current is present in resistor R1. The collector output potential is near the supply voltage, and transistor Q5 conducts. The positive voltage across the emitter resistor of Q5 rises to the high logic (1) state.

If one of the input gates is biased to a positive logic (1) state, its associated transistor conducts. The collector resistor R1 then reduces the collector voltage, and output transistor Q5 is biased off. Therefore, the output voltage is at the low logic (0) level. The circuit operates as a NOR gate.

Adders

In computers and other measuring instruments using digital ICs, various combinations of gates, inverters, and binary counters are used to perform the various mathematical operations. For example, it is possible to add digital numbers using just the logical states 1 and 0. The answer you obtain from the logic addition can then be converted to the decimal answer.

For example, decimal 4 can be added to decimal 3 in the following manner. Set down the four-bit binary values for 4 and 3.

$$
\begin{array}{ll}
0011 & (3) \\
0100 & (4)
\end{array}
$$

Note that the addition of these two quantities is 0111. From Table 6-1, it is seen that this is the binary equivalent of decimal 7.

Now, determine what happens when you try to add decimal 4 and decimal 5.

$$0100 \quad (4)$$
$$0101 \quad (5)$$

The addition of two logical 0's is always a logical 0, while the addition of a logical 0 and a logical 1 is always logical 1. However, in logic, the addition of two logical 1's is, of course, not 2 but 10. Hence, in the addition below, you will notice that a 10 has been placed under the one-plus-one addition column. As this column has a one-bit value, the 1 must be *carried over* into the column to the left. Therefore, as shown below, the sum is 1001 which in Table 6-1 corresponds to decimal 9.

$$
\begin{array}{cccc}
0 & 1 & 0 & 0 \quad (4) \\
0 & 1 & 0 & 1 \quad (5) \\
\hline
0 & 10 & 0 & 1 \\
\\
1 & 0 & 0 & 1 \quad (9)
\end{array}
$$

The addition of decimal 6 and decimal 7 is interesting.

$$
\begin{array}{cccc}
0 & 1 & 1 & 0 \quad (6) \\
0 & 1 & 1 & 1 \quad (7) \\
\hline
01010 & & & 1 \\
\\
1 & 1 & 0 & 1 \quad (13)
\end{array}
$$

In this example there were two carry-overs, and the final addition summation is 1101, corresponding to decimal 13.

Observe that everything is handled, including the shift from one column to the other, by the logic states of 0 and 1. Therefore, digital circuits suitably arranged can handle the complete process. Circuits that perform summations of this type are called adders.

Flip-Flops and Counters

Some popular digital IC circuits that are essential to computers are also widely used in test equipment associated with linear IC systems. These are the triggered and free-running types of multivibrators, plus frequency counters and dividers. The triggered type of multivibrator is commonly referred to as a flip-flop or toggle.

The basic flip-flop, shown as a discrete transistor circuit in Fig. 6-40, has two stable states. Basically, the circuit is a two-stage positive feedback arrangement with two outputs, Q and $\overline{Q}$. The two steady states are transistor Q1 off and transistor Q2 on, transistor Q1 on and Q2 off. A switch-over between states can be accomplished with the application of an appropriate voltage or pulse to the Set and Reset inputs of the flip-flop.

When a positive voltage is applied to the base of transistor Q1, the transistor is turned on. Its collector voltage drops and drives the base of transistor Q2 negative and into cutoff. Therefore, the output voltage is positive at Q and less positive at $\overline{Q}$.

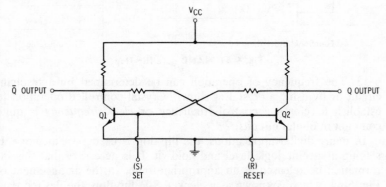

Fig. 6-40. Basic transistor flip-flop circuit.

A positive voltage applied to the base of transistor Q2 flips the operation back over. Transistor Q2 now conducts, and the negative collector voltage applied to the base of transistor Q1 drives Q1 to cutoff. Consequently, output Q is now *low* and output $\overline{Q}$ is *high*.

In computer vernacular, a logical 1 voltage at the Set input produces a logical 1 at output Q and a logical 0 at output $\overline{Q}$. Conversely, a logical 1 voltage at the Reset input produces a logical 0 output at Q and a logical 1 output at $\overline{Q}$.

Flip-flops of this type can be set up by connecting digital NAND and NOR gates in feedback configurations (Figs. 6-41 and 6-42). In each example, the device really has four definite states. However, the two important ones that set up the steady state logic conditions occur for Reset 0, Set 1 and Reset 1, Set 0. The outputs are as indicated by the truth tables.

As shown in Fig. 6-41B, the application of logical 1 voltages to both Reset and Set simultaneously causes *no change* in the output. In practical application, the Q and $\overline{Q}$ outputs are related inversely. It is also shown that the condition in which the logic inputs, Set and Reset, are both 0 is *not allowed*. A similar relationship exists for the NOR-gate flip-flop of Fig. 6-42.

The multivibrators shown are triggered types. However, with the proper selection of components and with the proper method of feedback, such a multivibrator can be made free-running or astable in its mode of operation. An example using two NOR gates is shown in Fig.

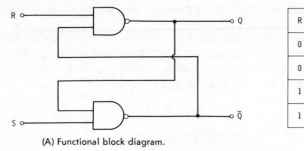

(A) Functional block diagram.

R	S	Q	$\bar{Q}$
0	0	NOT ALLOWED	
0	1	1	0
1	0	0	1
1	1	NO CHANGE	

(B) Truth table.

Fig. 6-41. NAND gate flip-flop.

6-43. The frequency of operation can be determined by a resonant circuit or by using a crystal as shown. Crystal-controlled operation to establish a reference control frequency or *clock frequency* is quite common in digital circuits.

In many digital applications, the flip-flop is used as a memory. It sets up a specific logic level and holds it. It is necessary that this information be released at an appropriate time. In the arrangement of Fig. 6-44, which is known as a clocked S-R flip-flop, the desired data

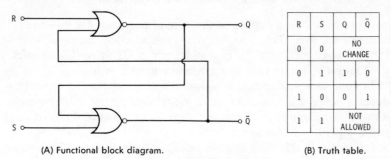

(A) Functional block diagram.

R	S	Q	$\bar{Q}$
0	0	NO CHANGE	
0	1	1	0
1	0	0	1
1	1	NOT ALLOWED	

(B) Truth table.

Fig. 6-42. NOR gate flip-flop.

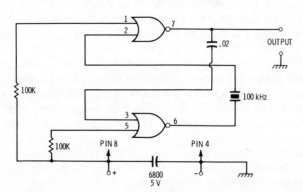

Fig. 6-43. Free-running crystal-controlled multivibrator using NOR gates.

226

is applied to the R and S inputs. However, the release of data to the output is under the control of the signal applied to the clock input.

Note that the conditions for R and S (0 and 1, 1 and 0) are the same as for the NAND-gate flip-flop discussed previously but the condition of $R = 0$, $S = 0$ differs. In this arrangement, output Q does not change but remains a logical 1 after the condition of $R = 0$, $S = 0$. Output is logical 0 after clocking by the $R = 1$, $S = 0$ condition. This indicates that the Q output is a logical 0 when the S data is a logical 0 regardless of whether the R input is logical 0 or logical 1. This possibility has advantages in counting and decoding circuits.

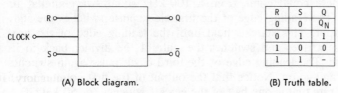

R	S	Q
0	0	Q_N
0	1	1
1	0	0
1	1	-

(A) Block diagram. (B) Truth table.

Fig. 6-44. Clocked S-R flip-flop.

Various types of flip-flop circuits can be employed as counters, singly or in combinations, to produce outputs that are some exact submultiple of the input clock frequency. They operate as frequency dividers by counting the input pulses and releasing a logic output after so many input pulses have arrived.

The Texas Instrument 7473 (Fig. 6-45) is an example of a TTL flip-flop that can be used as a counter. It is a dual device with two individual flip-flops that can be operated as a single two-to-one divider (Fig. 6-46), or by supplying the output of one to the input of the other as a four-to-one divider.

Counters can be cascaded to increase the total count. For example, three 2-to-1 dividers can be used to obtain a total count of 8-to-1 as

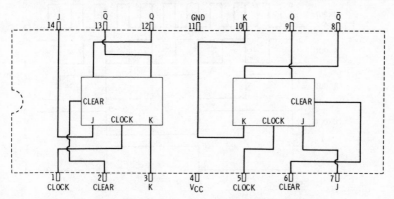

Fig. 6-45. The Texas Instrument 7473 TTL flip-flop.

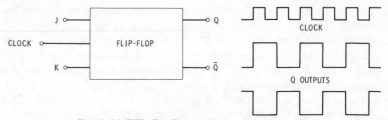

Fig. 6-46. TTL flip-flop used as a divider-counter.

shown in Fig. 6-47. The first waveform drawing shows the input clock pulses. These could come from a 100-kHz square-wave digital IC oscillator. The leading edge of the first clock pulse switches the state of the output. This state is held until the leading edge of the next clock pulse arrives and switches the state of the divider back to its initial logic. The leading edge of the third clock pulse again switches the state, and so on. Notice that the output of the first counter has a repetition rate that is one-half of the clock frequency, or 50 kHz.

The output of the first divider serves as the clock pulse for the second divider. Again the repetition rate is halved, and the output of the second divider would be 25 kHz. This division is repeated again by the last divider, producing a final output of 12.5 kHz.

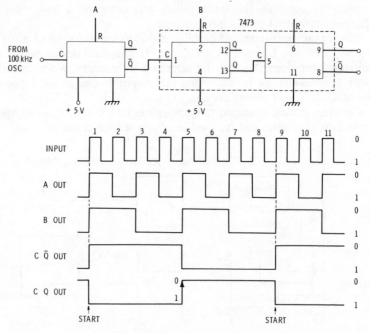

Fig. 6-47. An 8-to-1 counter.

An odd number count can be obtained by using a feedback path among the counters as shown by the five-to-one divider of Fig. 6-48. A capacitor provides the feedback path for a reset pulse that is connected from the Q output of the last counter to the Reset or Clear inputs of the first and second counters. This pulse is of a polarity that cancels out or inhibits the activities of the first two counters at the proper time to start a new five-pulse sequence.

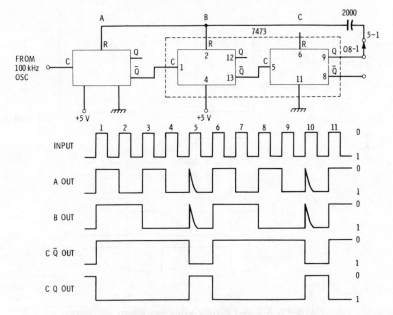

Fig. 6-48. A 5-to-1 counter.

In the waveforms of Fig. 6-47 it is to be noted that the edges of all five waveforms are coincident at the leading edge of each fifth clock pulse. Furthermore, the trailing edge of the Q output of the last counter is swinging from logical 1 to logical 0, a polarity that can be used to inhibit the first two counters. This activity coincides with the leading edge of the fifth clock pulse. The leading edge of the sixth clock pulse now switches all three counters as did the leading edge of the first clock pulse. The counters see everything as starting anew; they go off and try to count eight again, only to be met by another reset pulse coincident with the leading edge of the tenth clock pulse.

A feedback plan permits a three-to-one count using two 2-to-1 dividers connected in cascade (Fig. 6-49). Typical waveforms are shown, and the inhibit time is marked to show the point at which activities are reset.

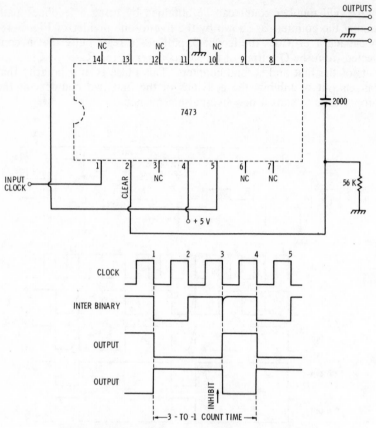

Fig. 6-49. A 3-to-1 divider using two binary flip-flops.

EXPERIMENT 5: DIGITAL DIVIDERS

General

The frequency divider changes an incoming frequency to a sub-multiple of that frequency. For example, when the input frequency is 100 kHz, a two-to-one divider produces a 50-kHz output frequency. Similarly, a five-to-one divider would change the 100-kHz input frequency to an output frequency of 20 kHz. In computer application, frequency dividers are referred to as counters because they produce one output pulse for every so many counted input pulses.

As covered in this chapter, two-to-one dividers can be connected in cascade to produce a desired overall even count. Also, with the use of feedback, such a cascade group can be made to operate as an odd-numbered counter. For example, two 2-to-1 counters in cascade produce an overall division of four. However, with the use of feedback the division can be three. Likewise, three 2-to-1 counters in cascade produce an overall count of 8. However, with feedback such a trio can be made to divide by 5.

A simple block with a clock input (feed) and one or two outputs (Q and/or $\overline{Q}$) symbolizes a basic two-to-one counter. Sometimes a set or reset input is also indicated for establishing odd counts and other functions.

Procedure 1: Two-to-One Dividers

The HEP-583 integrated circuit operates as a very simple two-to-one divider that can be supplied with signal from the output of the 100-kHz oscillator, as shown in Fig. 6-50. Drive to this flip-flop divider

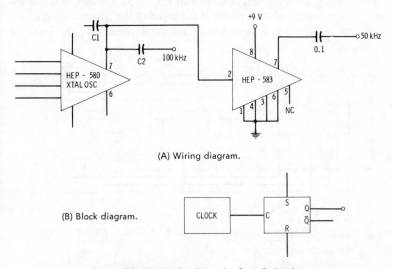

(A) Wiring diagram.

(B) Block diagram.

Fig. 6-50. Wiring hookup of a 2-to-1 divider.

is obtained directly by connecting terminal 7 of the oscillator directly to terminal 2 of the HEP-583. Output is taken off at terminal 7. As in the case of the HEP-580, the oscillator supply voltage is connected to terminal 8 while terminal 4 connects to common. Terminals 1, 3, and 6 also connect to common. The divider assembles very quickly.

Hook up the 9-volt battery to the oscillator and divider. Check the waveform at the output of the oscillator. Adjust the oscilloscope to display four square waves. Now transfer the oscilloscope to the output of the divider. You should observe two square waves instead of four, indicating a division by 2. Repeat the procedures of Experiment 4. You will now hear a distinct calibrated point every 50 kHz over the radio dial instead of at the previous 100-kHz points.

Procedure 2: Decade Divider

The 7490 IC is a decade divider in that it makes an overall division, or count, by a factor of 10. It includes four individual two-to-one counters mounted within a single dual in-line integrated-circuit package (Fig. 6-51). Counter A is an isolated two-to-one divider. Clock input is to terminal 14; the output is at terminal 12. Counters B, C, and D are connected in cascade with the necessary feedback to obtain a five-to-one division. The two-to-one and five-to-one dividers can be connected together externally to obtain an overall count of 10. The two sections may also be made to operate as separate two-to-one and five-to-one dividers. In our experiment, the two will be connected together externally to obtain an overall count of 10.

This wiring arrangement is shown in Fig. 6-51B. Note that the input is applied to terminal 1 which is the input of the five-to-one divider. Its output at terminal 8 is coupled back to terminal 14, the input of

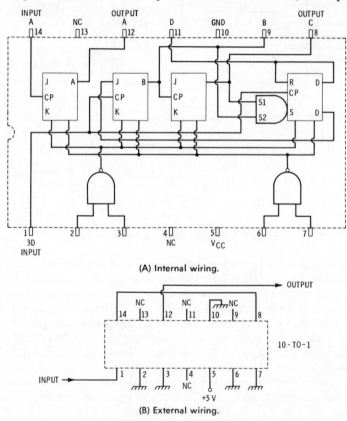

(A) Internal wiring.

(B) External wiring.

Fig. 6-51. Pin-out diagrams of the 7490 IC decade counter.

the two-to-one divider. The divide-by-ten signal is then removed at terminal 12. The complete wiring diagram is shown in Fig. 6-52.

Wire the circuit. Assuming that the oscillator frequency is 100 kHz, there are two outputs: one of 10 kHz and the other of 5 kHz. The output of the initial two-to-one divider is 50 kHz. After the five division of the decade counter, the output is 10 kHz (50/5). The final output from the last two-to-one counter is 5 kHz (10/2).

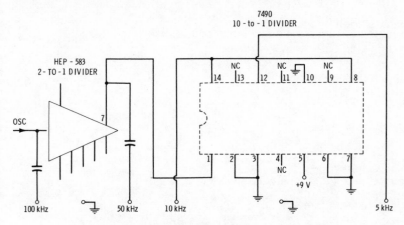

Fig. 6-52. Wiring hookup of the decade divider chain.

Apply power to the oscillator and the divider chain, and take a look at the square waves at the output of each terminal, observing the 100-kHz, 50-kHz, 10-kHz and 5-kHz outputs. You now have a complete divider chain making available these four calibrate frequencies. If you have precisely set the 100-kHz oscillator on frequency using the WWV signal, you will have a very accurate output at each of the four specified frequencies.

Turn on the a-m receiver, and couple the 10-kHz output to the antenna input system by wrapping a few turns of insulated wire around the staff antenna. Tune over the a-m radio dial. Notice that you will hear quite a number of heterodyning zero-beats as you cross the frequencies of the local broadcast stations. These channels are, of course, spaced exactly 10 kHz apart, and you will hear a zero-beat as your calibrated signal and incoming a-m signal beat together.

Turn on your shortwave or ham receiver. Switch on the bfo and note the calibration points every 10 kHz over the high-frequency dial. Try the 5-kHz output and you will note strong calibration points each 5 kHz. This is a help in locating and identifying the weak, long distance short-wave broadcast signals.

The five-stage divider. The divider takes a signal and gives it back at one-tenth its input. The complete wiring diagram is shown in Fig. 8-22.

Two bits are used. Assuming the 15-volt input (pin 9 is 100 Hz), there are two outputs: one of 10 kHz and the other at 1 kHz. There can also be bit lines two to be distributed. Since a dual two is used to provide all the decade chain, the output is 10 kHz (10,000). The output from the last five-stage divider is 1 kHz (1000).

Fig. 8-22. wiring diagram of the decade divider chain.

Apply power to the oscillator and the divider chain, and check each
of the seven outputs in succession of each terminal observation. In all
cases, SN7490, SN7400 and SN7420 chips. If a slow frequency output is
available, there is a simpler way to do so with almost impossibility as you
have observed that the 100 Hz output line is changing about once every
second, and will have every day or two unless a reasonable time may occur
to you. Frequencies.

The crystal oscillator, frequency and output the yield the output to the
dividing output system by connecting a few extra wires at logical levels around
the signal and counting the levels as a metallic divider whose final output is
the decade divider of introducing extra counts as you watch the
sequence of 15 volts in logic sequence. These should be no occasion,
however, as soon as the bits operation, it will reach a zero-beat counter,
calibrated output and measuring with a universal digital frequency meter.

From the oscillator or to handle frequencies with the oscillators, and
frequency distribution points every 10 seconds, begin to observe more.
Tie the 100 Hz output and you will increase the calibration once each
second. Hz this is a nice technique to a calibrator whenever, even the
same showing well spaced out.

7

Audio and AM / FM Systems

The home-entertainment electronics industry is the top user of special linear ICs. Specific devices have been designed for stage and system functions in audio amplifiers, radios, and television sets. Several typical and very specialized devices for application in audio systems and am/fm radio are detailed in this chapter. Internal and external circuits are discussed. At the end of the chapter a sequence of projects is started which culminates in the construction of complete fm stereo receivers and waveform generators.

AUDIO AMPLIFIERS

The *Basic Audio Stage*

Complete audio power amplifiers can be included in a single device. The Sprague ULX-2280 IC is an example (Fig. 7-1). This device is available in two configurations—the standard 14-lead dual inline, or a special 8-lead package that includes two heat sink tabs (GND connections of Fig. 7-1B). These tabs join to pins 3, 4, 5, 10, 11, and 12. They permit an external heat sink to be attached to the tab in a convenient manner while the printed wiring board is flow soldered.

A low-cost record player can be constructed with only a single device, and it will deliver a 3- to 5-watt output, using the heat-sink tab device (Fig. 7-2). Volume and tone controls are included. An optional series resistor and capacitor can be inserted if oscillation exists during certain load conditions. Overall amplifier gain is approximately 40 dB.

For stereo application two audio amplifiers can be included in the same device. One such unit, the Sprague ULN-2277, is shown in Fig.

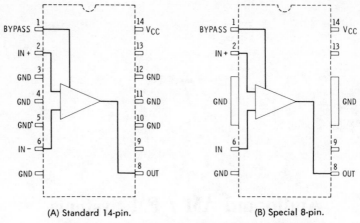

(A) Standard 14-pin.   (B) Special 8-pin.

Fig. 7-1. Pin-out diagram of Sprague ULX-2280 IC.

7-3. This linear IC is designed for use in stereo phonographs, am/fm stereo receivers, auto radios, tape players and recorders, intercom systems, and motion picture projectors. Two such devices can be the basic amplifiers of a quadraphonic sound system.

Using a supply source of 26 volts, the power output per channel is 2 watts. The schematic diagram of Fig. 7-4 shows that the input stage is a Darlington-configuration differential amplifier using transistors Q1 through Q4. Transistor Q5 serves as the differential load and supplies signal to the common-emitter power amplifier, transistor Q6. Level-setting is handled by transistor Q7, while the diode and associated circuits determine the idle current of the output transistors. A separate section contains a voltage regulator and ripple filter which supply the necessary voltages to the dual audio amplifiers.

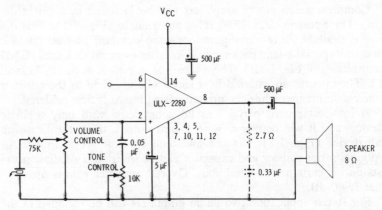

Fig. 7-2. A simple record player amplifier.

It should be noted that each channel also includes a feedback pin (terminals 7 and 8). As in operational amplifier application, a resistive voltage divider connected between the output, terminal 2, and the feedback input, terminal 7, can be used to control amplifier gain. This is shown in the schematic diagram of a low-cost stereo phonograph (Fig. 7-5). Note that feedback resistor R_f has a value of 100K and connects between terminals 2 and 7. Divider resistor R1 is selected so as to obtain the preferred gain. Typical values are given for gains between 28 and 46 dB.

As indicated, the component values for the second channel are identical to those of the first. Balance, tone, and volume controls are included.

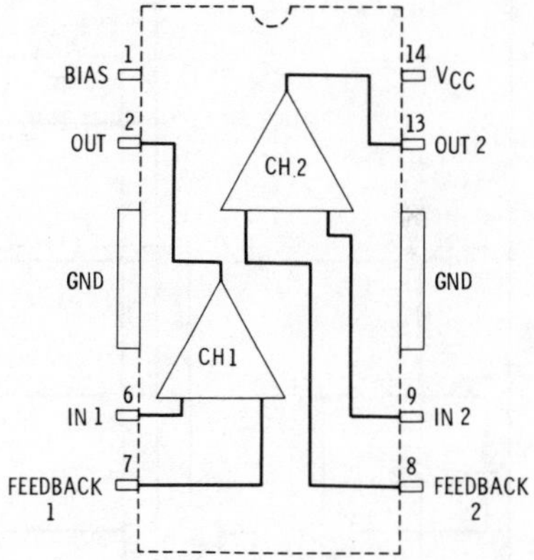

Fig. 7-3. Pin-out diagram of Sprague ULN-2277 IC.

A schematic for a complete two-channel 40-dB stereo power amplifier is given in Fig. 7-6. This arrangement assumes that the input signal is derived from a previous dual preamplifier to obtain higher sensitivity and a greater power output. A preamplifier that can be used in this application is the ULN-2126A integrated circuit (Fig. 7-7). The combination of two linear ICs used as a preamplifier and power amplifier permits the design of a quality stereo amplifier with 2-watts output per channel.

Schematic diagrams showing this application as a record player or stereo tape player are given in Fig. 7-8. The volume and tone controls as well as the balance adjustment are inserted in the coupling system between the two devices. Component values shown provide the proper equalization required in record and tape playback systems.

Hybrid Power Amplifier

The RCA HC1000 IC is capable of delivering 100 watts of audio into a 4-ohm load. Peak current demand is 7 amperes. Circuits for split-supply or positive-supply operation are given in Fig. 7-9. The

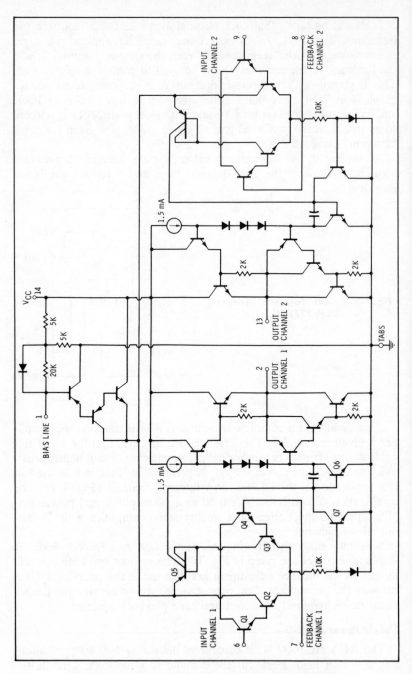

Fig. 7-4. Internal schematic of the Sprague ULN-2277 IC.

238

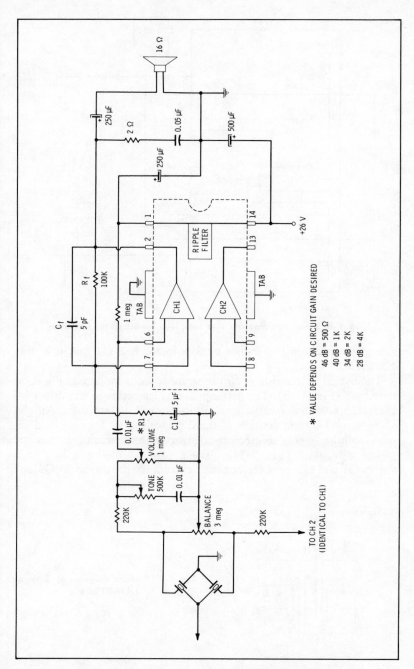

Fig. 7-5. Schematic diagram of a low-cost stereo phonograph.

239

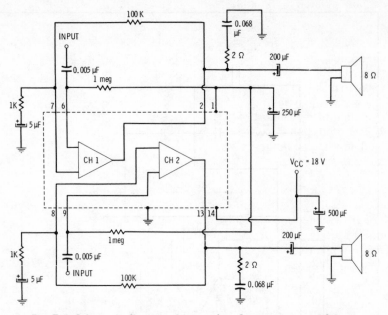

Fig. 7-6. Schematic diagram of a complete 2-watt stereo amplifier.

size of the package is 3 inches by 3½ inches and the weight is only 100 grams.

The hybrid IC consists of two separate sections mounted on a common base (Fig. 7-10). The preamplifier which serves as the driver includes 23 resistors, 7 capacitors, 6 diodes, and 9 transistors. An elevated second section consists of the 2 power output transistors, 2 diodes, and the associated mechanical structure and circuit.

The input stage (Fig. 7-11) consists of differential amplifier transistors Q1 and Q2, plus the constant-current source transistor Q3 and

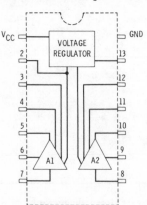

Fig. 7-7. Pin-out diagram of Sprague ULN-2126 IC.

associated diode. Output is derived across resistor R5 in the collector circuit of transistor Q1 and supplied to the base of the class-A amplifier, transistor Q5. Level setting is handled by the common-collector connection associated with transistors Q4 and Q5. The remaining stages are, of course, direct coupled, and a proper level set must be made to make certain that there is a low level dc output current. Proper idle current must also be established for correct minimum-distortion operation of the class-B output circuit.

The protection circuit within the dashed block protects the hybrid module when an improper load is present across the output. The output transistors consist of Darlington pairs Q8 through Q11.

The hybrid amplifier is made stable under inductive and capacitive loading by proper setting of the gain-bandwidth product and the use of the 10-μH suppression inductors and the series C7-R23 combination between terminals 4 and 5 (Fig. 7-11). This means that an output transformer can be used to step up or step down the load impedance that can be driven by the hybrid IC when feeding other than a 4-to-8-ohm load.

Two-Stage Stereo Preamplifier

The RCA CA3052 integrated circuit (Fig. 7-12) contains four identical and independent amplifiers. These can be connected into a dual-channel high-quality stereo preamplifier. The internal circuit is identical to that described in conjunction with Fig. 5-47; a complete schematic diagram of a single channel is given in Fig. 7-13. If external components are so selected that the amplifier gain is 46 dB (1000-hertz reference frequency) with a cartridge-delivered input signal of 5 millivolts, there will be a 1-volt output from the preamplifier.

The input stage should operate with a gain in excess of 40 dB to maintain a high signal-to-noise ratio. Since each amplifier has an open-loop gain of 58 dB, some stabilizing feedback can be incorporated. This feedback arrangement can also be used to establish the standard RIAA record equalization. Note in Fig. 7-13 that the output is transferred from pin 6 through a feedback and equalization network back to the inverting input of the amplifier at pin 7. The second amplifier operates with a flat response over the audio range and is rolled off at about 20 kHz by using capacitor C1 in the feedback path between pin 1 and pin 3.

Potentiometer R3 serves as a balance control between the two pre-amplifiers, with the 15-ohm resistor R5 connecting to exactly the same feedback tie point in the second channel. Total harmonic distortion of the preamplifier at 1-volt output is less than 0.3 percent.

The tone control is connected between stages and operates as a passive circuit. The values selected for this losser-type tone control produces a 20-dB loss at the reference frequency. An ultimate boost

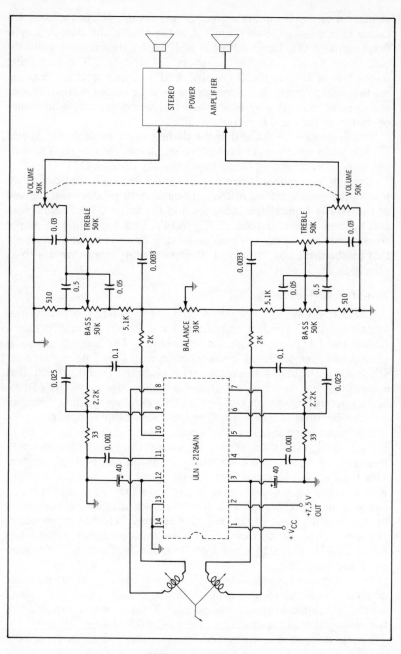

(A) Stereo record player.

Fig. 7-8. Linear IC

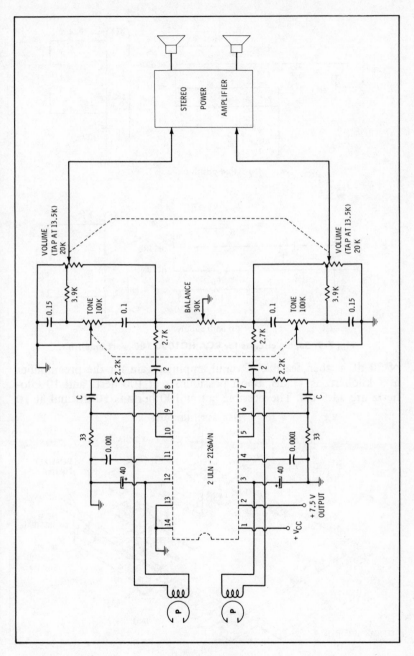

(B) Stereo tape player.

used as a preamplifier.

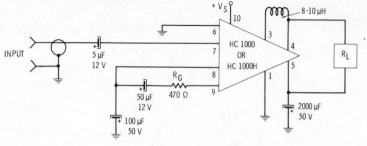

(A) Single positive supply.

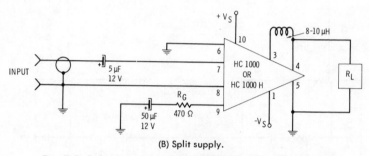

(B) Split supply.

Fig. 7-9. Supply circuits for RCA HC1000 100-watt amplifier.

of 20 dB is then feasible. Overall amplifier gain for the preamplifier at 1 kilohertz is 47 dB. Boost possibilities at 100 hertz and 10 kilohertz are 11.5 dB. The possible cut at 100 hertz is 10 dB and at 10

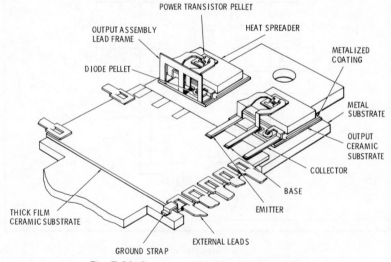

Fig. 7-10. Structure of high-power hybrid IC.

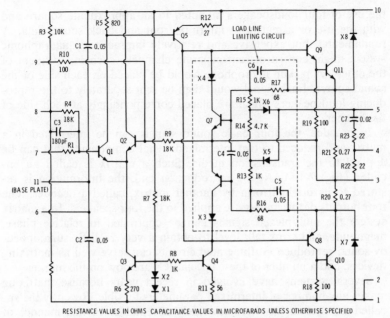

Fig. 7-11. Internal schematic of RCA HC1000 IC.

kilohertz is 9 dB. Volume level is adjusted at the input of the second stage.

Quadraphonic IC

Quadraphonic sound is a form of four-channel stereo. In this manner of stereo reproduction, the listener is surrounded by sound with

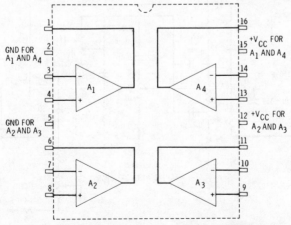

Fig. 7-12. Pin-out diagram of the RCA CA3052 IC.

the use of four loudspeakers mounted in an approximate square and with the use of a matrix to intermix front and back sound data. A fundamental, but expensive and relatively impractical, quadraphonic system might consist of four separate channels, each independent of the others. A pickup microphone would be placed on each side of the sound source. Each signal would then be sent separately to the reproducing loudspeakers which are placed correspondingly at each side of the listener.

In practice, the four basic sound signals can be intermixed in a matrix and reduced to two intermixed signal components that can be applied to the transmission media. Such a matrix is called an encoder (Fig. 7-14). At the reproduction end, the information is restored to its original form by another matrix, called a decoder. The four outputs of this matrix are applied to the four speakers. In a matrix system the multiplexed signals can be controlled in relative phase, magnitude, and cross-talk level to obtain a very realistic surrounded-by-sound reproduction. Integrated circuits can serve well as matrixing devices, and a number of these special ICs are now on the market.

Several systems have evolved in recent years because matrixing makes any number of intermixed parameters feasible. Recently the so-called SQ system has become widely acceptable. In this manner of encoding, the left and right channel signals have the following magnitude and makeup:

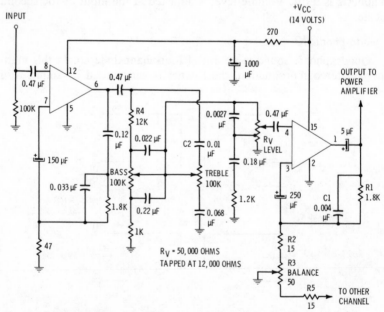

Fig. 7-13. Schematic diagram of one channel of the stereo preamplifier.

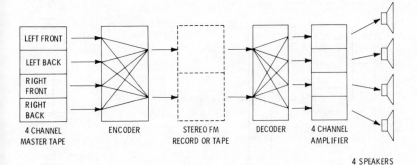

Fig. 7-14. Simple block diagram of a practical quadraphonic sound system.

$$L = L_f - 0.707jL_b + 0.707R_b$$
$$R = R_f - 0.707L_b + 0.707jR_b$$

where,
L_f is the left front magnitude,
R_f is the right front magnitude,
R_b is the right rear magnitude,
L_b is the left rear magnitude,
j is the phase difference of 90°.

The decoder plan for the SQ system is shown in Fig. 7-15. Such a simple decoder can be incorporated conveniently in an integrated circuit. Decoding is such that the signals applied to the front speakers are identical to the left and right channel signals. The two rear signals are derived in accordance with the phasing combination set up in the straight SQ decoder. Modifications to the reproduction can be made with the use of external resistors. These permit you to control the localization of sound sources in the listening area.

The circuit of the Motorola quadraphonic decoder MC1312 is shown in Fig. 7-16. The external resistor-capacitor network is in a Wien-bridge form. Values given provide a bandwidth between 100 hertz and 10 kilohertz, with the appropriate 90° phase difference. Supply voltage is connected to pin 12; pin 7 provides the common return. The right and left channel inputs are connected to pins 8 and 6, respectively. Front outputs appear at pins 2 and 11; rear outputs appear at pins 3 and 14.

In the basic SQ decoder, any sound reproduction intended for center front hearing results in equal outputs from all four speakers. This can interfere with center-front localization because the rear sounds are out of phase. This condition is avoided with some external blending of the two front and two rear outputs. An accepted blend combination is 10 percent between front speakers and 40 percent between rear

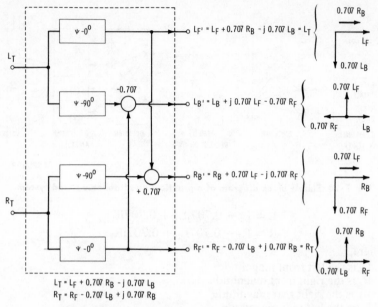

$L_T = L_F + 0.707 R_B - j\, 0.707 L_B$
$R_T = R_F - 0.707 L_B + j\, 0.707 R_B$

L_T AND R_T ARE COMPOSITE SIGNALS FROM SQ ENCODED RECORDS OR SQ BROADCAST.

Fig. 7-15. Plan of an SQ decoder.

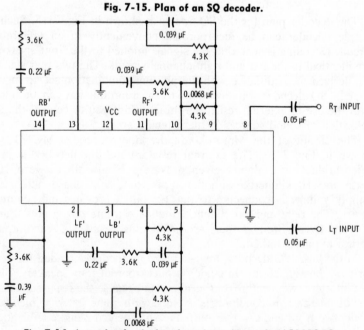

Fig. 7-16. A quadraphonic decoder using a Motorola MC1312 IC.

speakers. Such 10-40 blending or other blend combinations can be accomplished by connected appropriate external resistors between outputs, as shown in Fig. 7-17. In the example, a 10-percent front blend is obtained by using a resistor value of 47K for R1. A 7.5K value for resistor R2 provides the 40-percent blend between rear speakers.

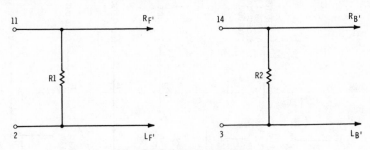

Fig. 7-17. Blending method.

A-M RADIO SUBSYSTEMS

ICs have been designed specifically for a-m radio application. These applications include mixers, local oscillators, i-f amplifiers, a-m detectors, and audio preamplifiers. A few include rf amplifiers as well.

The RCA CA3088 (Figs. 7-18 and 7-19) is an example. In addition to a converter (mixer-oscillator combination), a two-stage i-f amplifier, an a-m detector, and an audio preamplifier, it includes an output facility for a metering system, an i-f agc, and a delayed agc that can be applied to an external rf amplifier. The device operates from a 9-volt transistor radio battery.

Transistor Q1 functions as the converter. An appropriate external resonant circuit permits this manner of operation (Fig. 7-20). The small winding attached directly to the collector, pin 3, feeds back some of the rf energy into the oscillator tuned circuit that connects to the converter input, pin 2. Rf signal is applied to the same input.

The collector resonant circuit is in the form of a double-tuned transformer set to the i-f frequency. This transformer is located between pins 3 and 4; pin 4 is the input to the first i-f amplifier. The i-f stage is biased for 1 mA of current under zero-signal condition. Biasing is established by transistors Q2, Q3, Q4, and the four-diode biasing network (Fig. 7-19). Supply and biasing voltages are regulated by the internal zener diode Z1.

The second i-f amplifier and detector combination comprise transistors Q7, Q9, Q10, and Q12. Note that Q7 and Q10 are emitter followers, isolating the input and the output of the high-gain stage, transistor Q9. The operating point of the system is stabilized by the

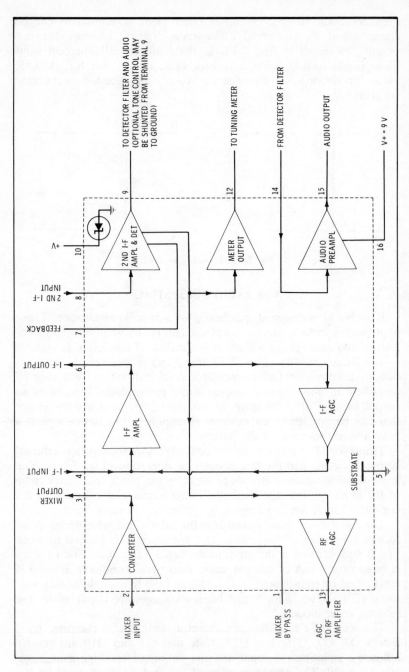

Fig. 7-18. Functional plan of RCA CA3088 IC a-m subsystem.

250

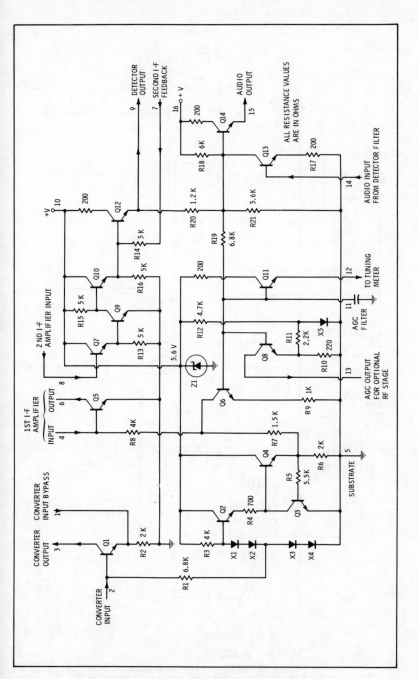

Fig. 7-19. Internal schematic of RCA CA3088 IC.

251

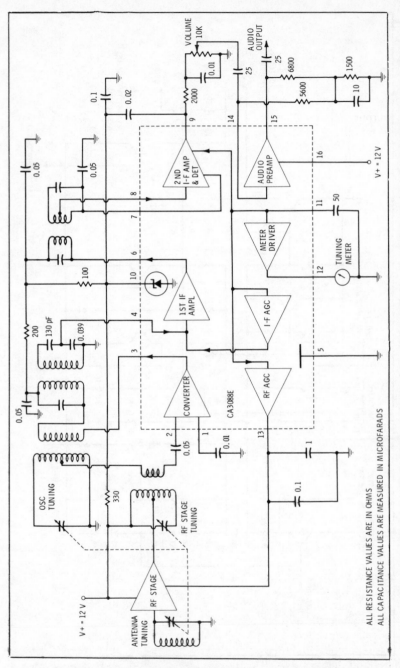

Fig. 7-20. External resonant circuit for a-m receiver using an IC subsystem.

252

external circuit that must provide a feedback path between the emitter of transistor 10 and the base of transistor Q7. This feedback path in the circuit of Fig. 7-20 is by way of the secondary winding of the double-tuned i-f transformer (pins 7 and 8) that joins the output of the first i-f amplifier to the input of the second i-f amplifier.

Transistor Q12 is connected as an emitter follower that is biased to have a quiescent current of 100 microamperes. The demodulated output is the rectified emitter current of this transistor. Demodulation is completed by attaching an appropriate resistor-capacitor filter to the detector output. This external filter and volume control can be seen attached to pin 9 of Fig. 7-20.

It is to be noted in Fig. 7-19 that a rectified component also develops a voltage at the junction of resistors R20 and R21. This dc component of voltage which is a function of the incoming signal carrier level is applied to the bases of the agc and metering amplifier, transistors Q6, Q8, and Q11. The emitter of transistor Q11 can be connected to an external tuning meter circuit. Agc voltage for an optional rf stage is made available at pin 13. The collector of transistor Q6 is the source of the agc bias for the first i-f amplifier.

Observe in Fig. 7-20 that the arm of the volume control applies signal to pin 14 which is the input of the audio preamplifier. Amplified audio for driving a succeeding audio power output stage or IC is removed at the emitter of the emitter-follower output stage, transistor Q14. Transistor Q13 operates as an isolation gain stage.

The Sprague ULX-2137 IC includes an rf amplifier stage in its subsystem integrated circuit (Fig. 7-21). However, it does not incorporate a detector and audio preamplifier. This IC also has been designed mainly for a-m radio application, and especially for use in auto radios (Fig. 7-21B).

Note that very few external components are required except for a resonant tuning circuit and interstage i-f transformer. Subsystems of this type typify the design of modern a-m radios. They permit the construction of compact and low-power units.

FM RADIO ICs

A combination of radio-frequency amplifier and i-f detector subsystem permits the design of an fm receiver using four to six integrated circuits. A popular rf amplifier is the RCA CA3005 (Fig. 7-22). It consists of a differential amplifier, a constant-current transistor source, and a temperature compensating diode. These can be interconnected to operate either as a straight rf amplifier or as a frequency converter.

A practical fm tuner using two ICs employs one as an rf amplifier and the second as a converter (Fig. 7-23). The incoming fm sig-

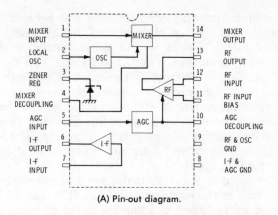

(A) Pin-out diagram.

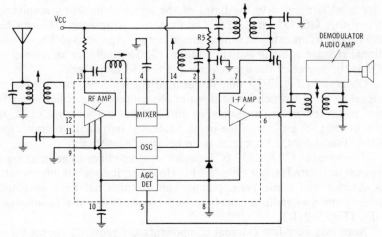

(B) Auto-radio application.

Fig. 7-21. Sprague ULX-2137 IC subsystem and auto radio application.

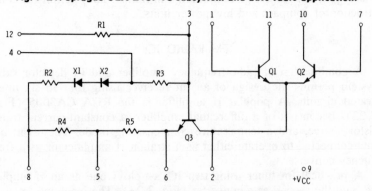

Fig. 7-22. Internal schematic of RCA CA3005 IC.

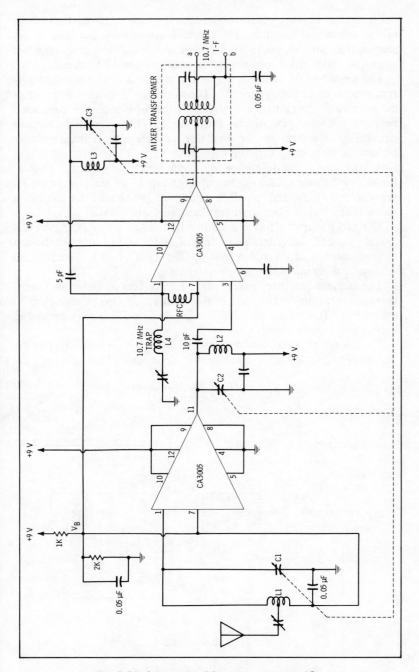

Fig. 7-23. Schematic of fm tuner using two ICs.

nal is applied to pin 1 which connects to the base of the first transistor of the differential amplifier. Pin 7, which connects to the base of the second transistor, operates at common potential. Output is removed at pin 11 and is developed across the parallel-tuned LC circuit.

The amplified fm signal is applied to the base of the constant-current source of the converter IC. The converter oscillates because of the positive feedback path introduced by the 5-pF capacitor connected between collector pin 10 and base pin 1. The associated resonant circuit L3-C3 determines the oscillator frequency. Note that the tuner includes a three-gang variable capacitor.

A 10.7-megahertz difference frequency is developed in the double-tuned transformer connected to collector pin 11 of the converter. To prevent instability and possible oscillation at the i-f frequency, a series-tuned trap is connected between pin 1 and ground.

The tuner output is fed to a similar IC used as an i-f amplifier, followed by an IC subsystem consisting of additional i-f amplification, a discriminator, and an audio preamplifier (Fig. 7-24). The internal schematic diagram of the fm subsystem is given in Fig. 7-25.

The sequence of three differential i-f amplifiers in series also functions as an fm limiter; limiting begins with an applied i-f signal of 30 microvolts. Total gain of the i-f subsystem is 70 dB as measured between terminals 1 and 5.

The fm signal at terminal 5 is applied to the primary winding of the phase-shift discriminator transformer. The secondary winding de-

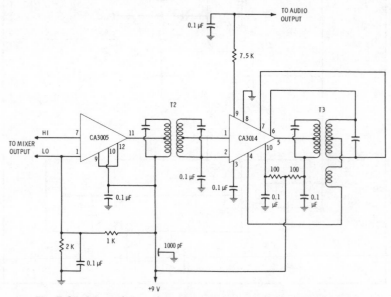

Fig. 7-24. I-f amplifier, phase discriminator, and audio preamplifier.

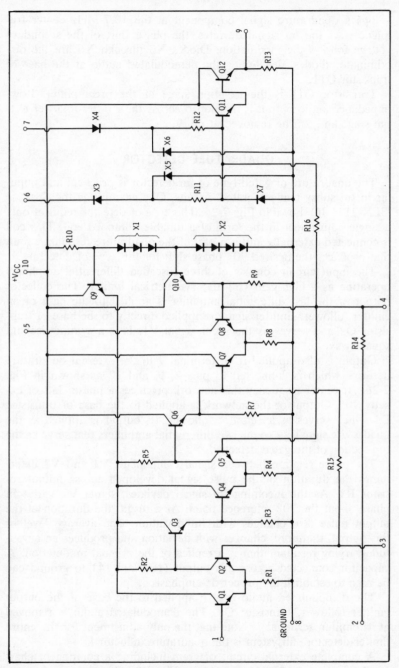

Fig. 7-25. The RCA 3014 IC fm subsystem.

257

velops a quadrature signal component at the 10.7-MHz center frequency. As the fm signal deviates, the phase shift of the secondary voltage follows the modulation. Diodes X3 through X6 are the discriminator diodes, developing the demodulated audio at the base of transistor Q11.

Transistor Q11 is the isolation stage of the preamplifier. Low-impedance audio output is developed across the emitter-follower output stage and can be removed at pin 9.

QUADRATURE DETECTOR

The quadrature or gated-type of fm detector is practical and popular in fm subsystem integrated circuits. One example is the Sprague ULN-2111 IC shown in Fig. 7-26. This type of detector requires only a single adjustment in the form of a tunable untapped coil. This coil is connected externally and is a part of the quadrature resonant circuit that produces the desired 90° phase shift for this type of detector.

The input circuit consists of three cascaded differential amplifiers operating as a high-gain (60-dB) symmetrical limiter. The collector output of the last differential amplifier is applied to the base of an emitter follower. Emitter signal is applied directly to the base of transistor Q2 and serves as the input signal V1. It is a symmetrical fm square wave.

Output is also applied from terminal 9 to the external quadrature network which is connected to pins 2, 9, and 12 as shown in Fig. 7-26. At center frequency this network produces a phase shift of exactly 90°. Output of the network is applied to the base of transistor Q1 which serves as an emitter follower. Its output is applied as the quadrature signal V2 to the two differential amplifiers that serve as the top section of the gated detector.

The phase relation between signal components V1 and V2 determines the duration of the pulse signal developed across output resistor R1. As the incoming fm signal deviates, signal V2 varies in phase about the 90° reference point. As a result, the duration of the output pulse also changes with fm deviation. The average level of this output, therefore, changes with deviation and produces an amplitude-varying resultant that is a replica of the original modulation. A capacitor connected across this output (terminal 14) to ground can be used to establish the proper de-emphasis.

The demodulated audio is also applied to the base of the output emitter follower, transistor Q8. The demodulated audio is removed at its emitter, terminal 1. Note that the only adjustment for the entire limiter-detector subsystem is the quadrature inductor L.

A complete external circuit diagram including a component chart for the Sprague ULN-2111 is given in Fig. 7-27. This information can

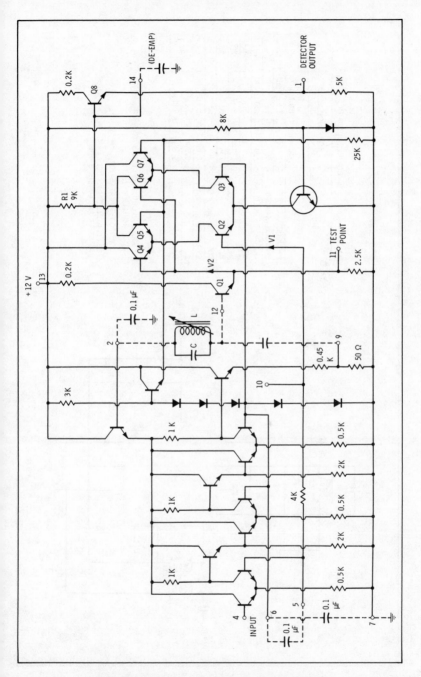

Fig. 7-26. Sprague ULN-2111 fm subsystem IC.

259

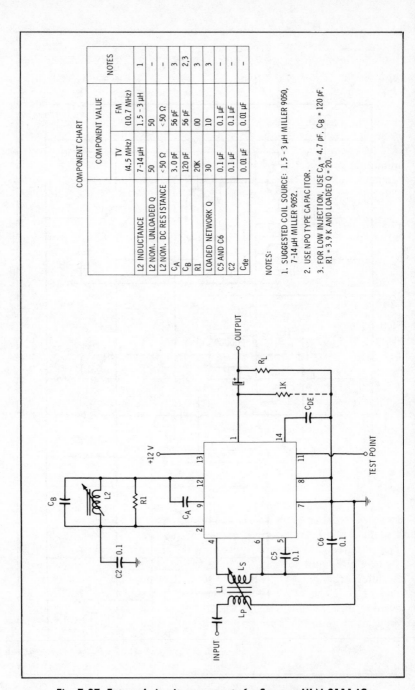

COMPONENT CHART

	COMPONENT VALUE		NOTES
	TV (4.5 MHz)	FM (10.7 MHz)	
L2 INDUCTANCE	7-14 µH	1.5 - 3 µH	1
L2 NOM. UNLOADED Q	50	50	-
L2 NOM. DC RESISTANCE	<50 Ω	<50 Ω	-
C_A	3.0 pF	56 pF	3
C_B	120 pF	56 pF	2,3
R1	20K	00	3
LOADED NETWORK Q	30	10	3
C5 AND C6	0.1 µF	0.1 µF	-
C2	0.1 µF	0.1 µF	-
C_{de}	0.01 µF	0.01 µF	-

NOTES:

1. SUGGESTED COIL SOURCE: 1.5 - 3 µH MILLER 9050, 7-14 µH MILLER 9052.

2. USE NPO TYPE CAPACITOR.

3. FOR LOW INJECTION, USE C_A = 4.7 pF, C_B = 120 pF. R1 = 3.9 K AND LOADED Q = 20.

Fig. 7-27. External circuit components for Sprague ULN-2111 IC.

be used to apply the subsystem in an fm receiver or as the sound system of a television receiver. In tuning the subsystem, one need only apply a 5-millivolt fm modulated signal to pin 4. The inductor is then adjusted for maximum recovered audio at pin 1, or maximum rf voltage at test pin 11.

More elaborate fm subsystems are available for use in high-performance and/or communication receivers. Such devices have great limiting sensitivity and interference rejection, and very low distortion. Their design provides for a maximum of interstation noise rejection, a minimum of response off on the sides, a facility for driving a signal-strength indicator and tuning indicator, a stereo enabling voltage, a delayed agc, and a versatile afc capability.

The system block of an RCA CA3089 IC shows the general plan of such an elaborate fm subsystem (Fig. 7-28). The basic stages consist of the three symmetrical limiters and a quadrature detector. The detector makes available a push-pull output, and therefore separate components are available for audio and afc.

There is a squelch detector that reacts to the signal-to-noise ratio in the limited signal. Output of the squelch detector is then used to obtain noise suppression when tuning between stations, as well as to reduce side tuning responses. The squelch threshold control is attached externally between terminals 5 and 12.

The integrated circuit includes an elaborate level-detecting circuit which sums the various levels in an adder. The combined level is applied to the tuning meter as well as to the stereo disable arrangement for the multiplex decoder. The level-detector circuit associated with the first limiter is also used to develop delayed agc for the tuner.

STEREO DECODERS

The spectrum distribution for the stereo multiplex broadcast signal is shown in Fig. 7-29. Left- and right-channel signals in proper proportion must be transmitted and then demodulated and decoded so as to recover the original two-channel audio information. A sum signal (summation of left- and right-channel signals) is transmitted and occupies a frequency spectrum up to 15 kHz. This portion of the complete signal can be demodulated by a monaural receiver in normal fashion. A difference signal $(L - R)$ is also transmitted. When the $(L + R)$ and $(L - R)$ signals are recombined in the matrix of the decoding system, the original left (L) and right (R) audio signals are recovered at the output. These are applied to a stereo audio amplifier and output.

The $(L - R)$ signal is transmitted on a 38-kHz subcarrier frequency. A double-sideband method of transmission using carrier suppression is used. Signal is known as the $L - R$ (dsb) component.

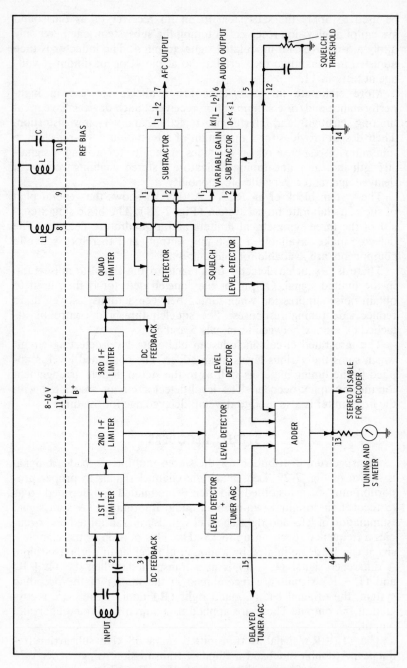

Fig. 7-28. Functional plan of RCA 3089 fm IC.

262

The third signal conveyed is a 19-kHz pilot subcarrier. This very-low-amplitude component is used in the decoder to regenerate the 38-kHz subcarrier for proper demodulation of the $(L - R)$ information.

In an fm stereo broadcast receiver the fm multiplex decoder is located after the fm detector. The complete decoding of the fm multiplex signal can be handled by a single integrated circuit.

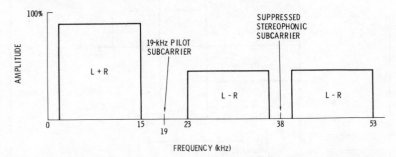

Fig. 7-29. Spectrum distribution for stereo multiplex broadcast signal.

An example of such a circuit is the Sprague ULN-2122 IC (Fig. 7-30) which is driven by the composite signal derived at the output of the standard fm detector. In fact, this multiplex decoder can follow right after the output of the IC fm subsystem of Fig. 7-27. Its output will be the original left- and right-channel audio signals.

The simplified diagram of its internal circuitry is given in Fig. 7-31. The composite input signal at transistor Q4 follows two paths. One of these is through a chain of amplifiers that applies the signal in composite form to the base of transistor Q22, which is one of the lower differential pair of the matrixing decoder. To obtain ideal bias matching, there is a similar dc path from the input through the bias matching transistors to the base of transistor Q19. This connection ensures a properly matched decoder and clean left- and right-channel outputs. Good balance is obtained, and that balance becomes independent of common-mode supply-voltage changes and temperature variations. It also aids in attenuating noise under weak receiving conditions, particularly in the pilot chain.

A low-level antiphase signal is also present, and its level is set by the ratio of resistor R24 to R40. This facility compensates for errors in the multiplex signal process. Control R40 is adjusted for the very best separation between the left and right output channels.

The pilot signal is applied to a 10-kHz pilot frequency amplifier. The amplified 19-kHz component is then applied to a frequency doubler comprised of transistors Q12, Q14, and Q16. In effect, the rectifier action of the Q12-Q14 combination produces a 19-kHz pulse

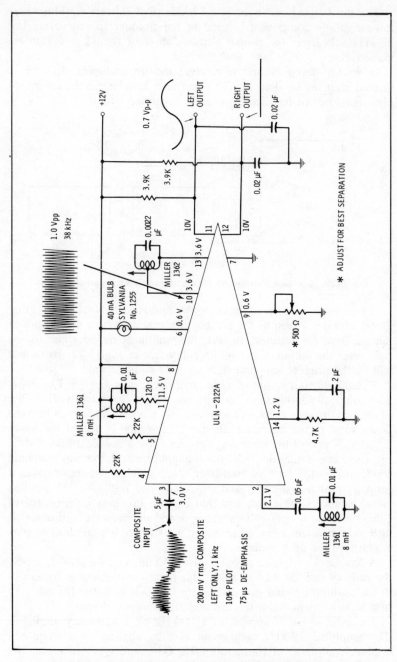

Fig. 7-30. External schematic of decoder using Sprague ULN-2122 IC.

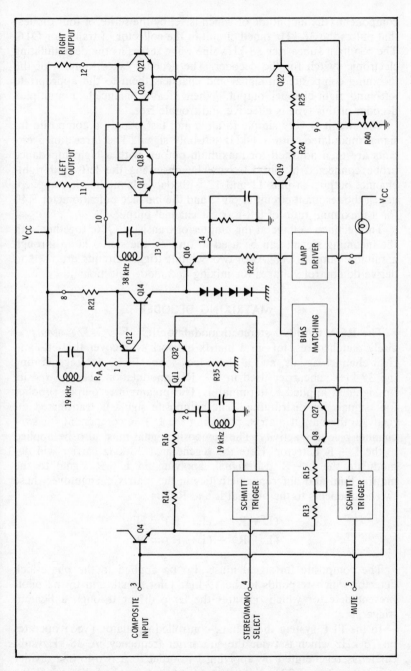

Fig. 7-31. Simplified internal schematic of Sprague ULN-2122 IC.

component (the amplitude of which is set by the series of four diodes) that pulses the 38-kHz tuned circuit in the collector of transistor Q16. The resultant subcarrier 38-kHz sine wave serves as the demodulating electronic switch for the decoder. The decoder takes samples of the incoming composite fm signal and channels them to the appropriate left- and right-channel output. When ideally balanced, a multiplex decoder of this type is effective and trouble free.

The circuit is very simple to align and test. Apply a composite fm signal modulated with a 1-kHz left-only signal. The three tuned circuits are then adjusted for maximum output, with a high impedance probe connected at pin 13. Now by observing the left- and right-channel outputs at pins 11 and 12 with only left modulation applied, adjust the resonant circuit at pin 2 and the balance potentiometer R40 for a maximum ratio of left-to-right channel output.

The rectified voltage at the emitter of transistor Q16, together with the incoming signal, can be used to turn on the stereo lamp through a suitable lamp driver. Also, two Schmitt trigger circuits are used to derive dc control voltages for mixing and mode selection.

MATRIXING DECODER

The RCA CA3090 stereo demodulator IC (Fig. 7-32) uses precisely matched resistors in a matrix network to recover the left- and right-channel signals, and a phase-lock loop (PLL) for regenerating the 38-kHz subcarrier used in $L - R$ demodulation. The composite fm signal is applied to terminal 1. The preamplifier output supplies four signals for distribution. The composite signal is transferred directly to the output matrix, providing the $L + R$ component for two-channel recovery activity. The composite signal must also be applied to the $L - R$ detector. Here the regenerated 38-kHz carrier will demodulate the $L - R$ dsb signal, applying an $L - R$ signal to the matrix. Sum and difference activities in the matrix change over these two components to the original L and R data:

$$(L + R) + (L - R) = 2L$$
$$(L + R) - (L - R) = 2R$$

The composite fm signal must also be applied to the phase-lock detector which responds to the 19-kHz pilot signal, and to the pilot-present detector which operates the lamp driver through a Schmitt trigger.

In the PLL system the voltage-controlled oscillator (vco) operates on 76 kHz which is twice the subcarrier frequency. A 38-kHz subcarrier is regenerated by applying its output to a two-to-one divider (top left of Fig. 7-32). Note that its output is applied directly to the

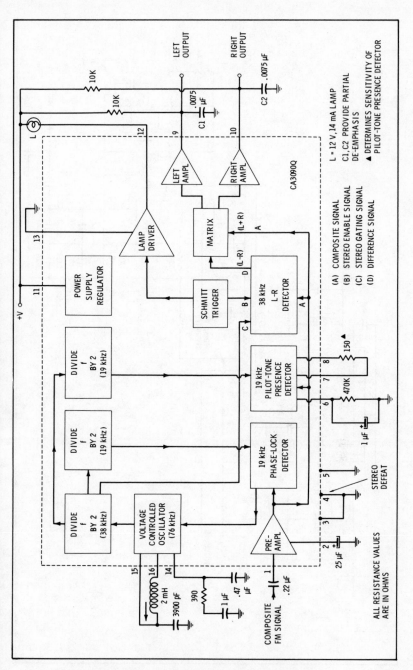

Fig. 7-32. RCA CA3090 IC stereo decoder.

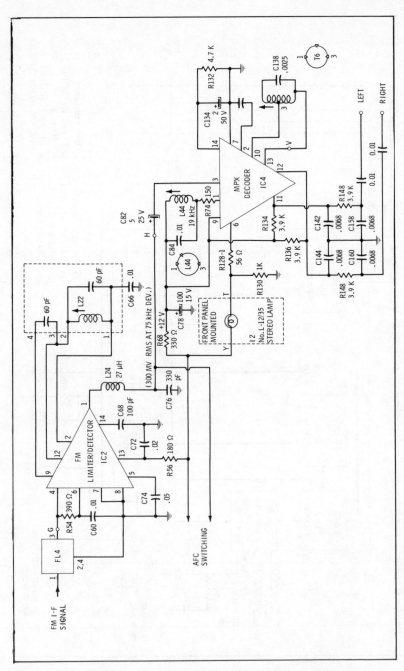

Fig. 7-33. Fm detector and decoder section of an am/fm receiver.

L − R detector. This detector is activated whenever a pilot tone is present and operates the Schmitt trigger.

There is a further division by two to produce a 19-kHz component that is applied to the phase-lock detector. Here it is compared with the incoming 19-kHz pilot frequency. A phase lock-in results and the 76-kHz vco is tied in frequency and phase to the incoming pilot component.

There is a second two-to-one divider that drops the 38 kHz to 19 kHz for application to the pilot-presence detector.

The limiter detector and multiplex decoder are a popular pair in many am/fm stereo receivers. Such a combination (Fig. 7-33) is used in a Sylvania model high-fidelity stereo console. The location of these two ICs is shown in Fig. 7-34.

The i-f signal is applied to the input of the IC by way of ceramic filter FL4. The fm limiter and detector IC has but one tuning adjustment, the quadrature coil. This coil can be seen just below the IC2 socket in Fig. 7-34.

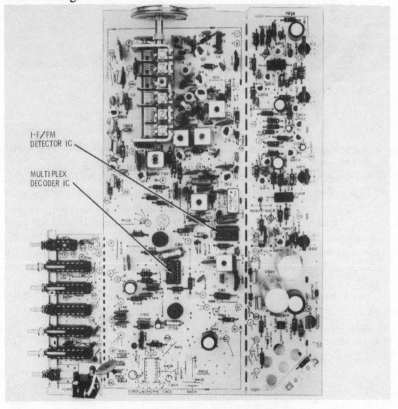

Fig. 7-34. Sylvania am/fm radio chassis.

The composite fm signal is applied to the multiplex decoder at pin 3. The decoder has two adjustments, 19 kHz and 38 kHz, mounted above and beneath IC4 (Fig. 7-34). Left- and right-channel outputs are supplied to a succeeding stereo audio amplifier.

PROJECT 1: AUDIO POWER AMPLIFIER IC

General

The Sinclair IC-12 audio IC is an excellent device for demonstrating the capability of integrated circuits in audio. An integral part of this 5-watt audio output device is its permanently attached heat fins (Fig. 7-35). A prepared printed-circuit board can be purchased with the IC, permitting the construction of a basic audio power amplifier. A current parts list is supplied with the kit by the manufacturer. Optimum supply voltage falls between 18-28 volts, although the unit functions with as little as 8 volts.

Fig. 7-35. Sinclair IC-12 audio IC.

Using a supply of 28 volts, the power output into 8 ohms is 6 watts rms. The unit will supply an output to load impedance values of between 3 and 15 ohms. This level of power output can be obtained with an input of only 30 millivolts. Other characteristics are a voltage gain of 250, input impedance of 250K, and a zero-signal current demand of only 10 mA. The power output stages operate class B, and the supply current demand increases with the level of the audio input signal. The total harmonic distortion is typically 0.1 percent, the signal-noise ratio is approximately 70 dB, and the frequency response extends between 15 Hz and 500 kHz.

The schematic diagram of the project is given in Fig. 7-36. The circuit is constructed on the supplied printed-circuit board (Fig. 7-37). A gain control potentiometer can be connected across the input (between terminals B and C) to permit control of the volume level. The functions of the various components are as follow:

C1 —The value of input coupling capacitor C1 is not critical, although a low value can be used to control the base rolloff response.

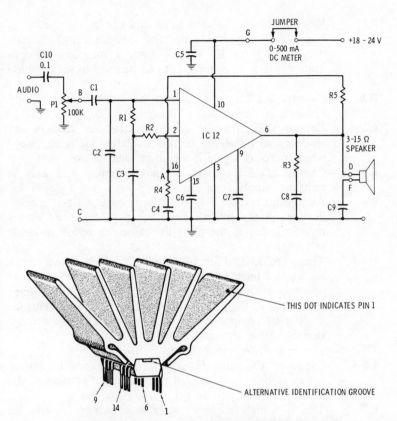

Fig. 7-36. Schematic diagram and pin read of IC-12 circuit.

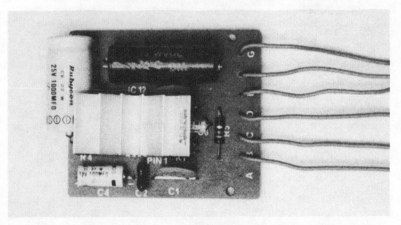

Fig. 7-37. Audio module with IC-12 and other components mounted on circuit board.

R1 —Resistor R1 can be used to control the input impedance. For a very high value of input impedance, the ohmic value of resistor R1 can be made 1 megohm. In conjunction with capacitor C1 it can be used to set the base rolloff.

R2 —Resistor R2 helps to hold up the input impedance and operates in conjunction with capacitor C3.

C2 —Capacitor C2 is a stabilizing capacitor that reduces any tendency to instability or self-oscillation. It is necessary when source of signal is a high-impedance one.

C3 —Capacitor C3 improves amplifier stability and aids in reducing supply ripple.

C4 —Along with resistor R4, this capacitor has a significant influence on the overall low-frequency response. Values recommended in the parts list cause the rolloff to occur at 18 hertz.

R4-R5—These are the two feedback resistors which connect from the output back to the inverting input of the device in operational-amplifier fashion. Actually the output voltage approximates the ratio of $(R4 + R5)/R4$. The influence of internal components is such that an effective voltage gain of 250 is obtained.

C5 —This is a power-supply filter.

C6 —Capacitor C6 maintains high-frequency stability, but at the same time it is selected to minimize distortion at the highest frequency to be conveyed.

C7 —This is a high-frequency stability capacitor, usually for fast negative-going sweeps of the input signal.

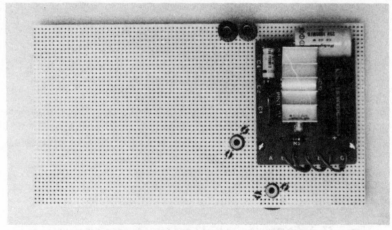

Fig. 7-38. Audio module attached to vector board.

C8-R3—These compensate for reactive components in the load (loudspeaker inductance).

C9 —Capacitor C9 influences the bass-frequency response of the amplifier.

Construction

Assemble the amplifier on its supplied printed-circuit board. By using suitable spacers, the audio module can then be fastened to a larger vector board (Fig. 7-38). A volume control and a phono-input receptacle can be attached to it. In Project 2 a small IC receiver will be attached to the same vector board and will be used to supply signal to the audio module.

Operation

Connect a 3- to 16-ohm speaker across the output. The ideal match is obtained with an 8-ohm speaker. Use an 18- to 24-volt supply capable of delivering a peak current of at least 300 mA. Audio signal is supplied through isolating capacitor C10 and a 100K volume control (Fig. 7-36).

The audio source can be a record player or a tape player. Connect its output across the audio input jack. Turn on the player and audio module and listen to the output. The music quality is good.

Insert a 0- to 500-mA dc meter in the (+) supply line (Fig. 7-36). Note that with no audio input the meter reading is very low. It rises to a peak value on a strong audio passage.

Connect a high-output, high-impedance microphone to the input. Position the microphone or speaker in an adjacent room to prevent feedback howl. Note the strong audio output that permits the integrated-circuit audio module to operate as a paging or public-address audio amplifier.

Attach an oscilloscope across the speaker. Connect an audio generator to the input. Set the volume control to the midposition. Adjust the oscilloscope for proper operation. Turn on the audio generator and set its frequency to 1000 hertz. Adjust the audio output level of the generator and the volume control until maximum undistorted output is obtained. Under this condition the dc current drawn by the audio module should be in the 200- to 300-mA range.

Check the output level at 100 hertz and 10,000 hertz. Vary the audio oscillator frequency above 10,000 hertz and note that above 25,000 hertz the response begins to roll off. Likewise there will be a gradual decline in output as the audio frequency is decreased below 100 hertz and down past 40 hertz. The good high-level output and low distortion of the amplifier are worthy of note. In the next project the amplifier will be used in conjunction with a simple integrated circuit radio broadcast receiver.

PROJECT 2: A-M RADIO

General

A single-chip integrated-circuit a-m radio is investigated in this project. Initial information about the Ferranti ZN414 IC was given in Chapter 6. It consists of a ten-transistor tuned radio-frequency amplifier including a transistor detector (Fig. 7-39). There are three radio-frequency stages and a high-impedance input stage. Agc action permits the device to operate with a wide range of input signal levels.

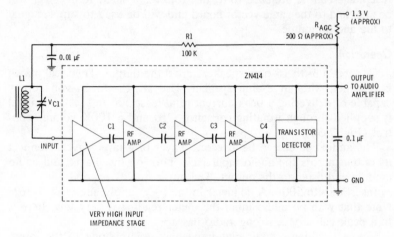

Fig. 7-39. The ZN414 IC circuit plan.

Externally a high-Q resonant circuit is needed and must be tunable over the desired input frequency range. The device itself performs well between 150 kHz and 3 MHz. Agc action and the feedback path is set by resistors R1 and R_{agc}. Supply voltage to the device must be 1.2 to 1.6 volts. If the device is to operate from a 9-volt battery, you can use the resistive divider shown in Fig. 7-40.

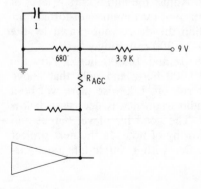

Fig. 7-40. Circuit for 9-volt operation.

274

Construction

The ZN414 can be used with the audio module of Project 1 to obtain a high-level audio drive for a speaker. The circuit arrangement (Fig. 7-41) includes two supply sources—a small battery for the radio and the previous source for the audio module. The receiver can be mounted on the same vector board (Fig. 7-42).

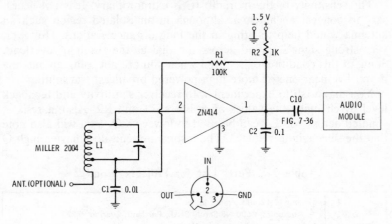

Fig. 7-41. Radio circuit.

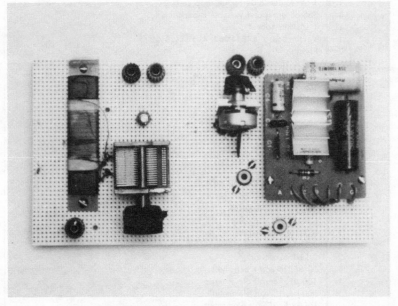

Fig. 7-42. IC detector mounted with audio amplifier.

Inductor L1 is a Miller ferrite loop mounted on a flat strip (Fig. 7-42). The Q of this coil at 790 kHz approaches 500. The high Q provides good selectivity for the tuned rf receiver. The output level is high enough to be applied directly to the input of the audio module through capacitor C10 of Fig. 7-36.

Operation

The sensitivity of the trf radio IC is extraordinary. It is not necessary to connect an antenna, although in an isolated region such an antenna would help to bring in the long distance stations. However, very strong signals spread across the dial as the result of overload. Some of this condition exists even when you are not using an antenna if you live near one or more high-powered broadcast transmitters.

You may wish to experiment with receiver sensitivity and feedback level by changing the values of resistors R1 and R2. Also, a resistor shunted across coil L1 can be used to lower its Q. You will also note that the selectivity is reduced. The resonant circuit does have a high Q

Table 7-1. Parts List for Projects 1 and 2

1	Sinclair IC-12 audio module
	(Audionics, 8600 NE Sandy Blvd., Portland, Oregon 97220)
1	Ferranti, a-m radio module ZN414
	(Ferranti Electric, East Bethpage Rd., Plainview, N.Y. 11803)
1	J. W. Miller 2004 ferrite loop coil
1	365-pF broadcast variable capacitor
1	Transistor socket
1	Vector board (84P44-062) 4½″ by 8½″
2	Phono sockets
1	8-ohm speaker
1	24-volt 300-mA power source
1	1½-volt D battery
5	Binding posts
1	100K potentiometer
1	18-ohm, ½-W resistor
1	100-ohm, ½-W resistor
1	1000-ohm, ½-W resistor
1	27K, ½-W resistor
1	100K, ½-W resistor
1	150K, ½-W resistor
1	270K, ½-W resistor
1	100-pF disc capacitor
1	500-pF disc capacitor
1	0.001-μF capacitor
2	0.01-μF capacitor
3	0.1-μF capacitor
1	10-μF, 15-volt electrolytic
1	100-μF, 15-volt electrolytic
1	500-μF, 15-volt electrolytic
1	1000-μF, 25-volt electrolytic

and there is some tendency to self-oscillation. If so, it can be reduced by shunting a 100K resistor across the coil.

This little receiver is capable of driving quite a large speaker. More than enough volume level is present even for the weaker incoming signals.

8

ICs in Television Receivers

Most of the functional sections of black and white and color television receivers are adapted to integrated circuit use. The application of ICs in all sections is a possibility within the next several years.

FUNCTIONAL SECTIONS OF A TELEVISION RECEIVER

Although an integrated circuit tuner is feasible, it is not common in today's models. Usually a discrete bipolar FET or, more often, a hybrid FET-bipolar combination is used. Video i-f integrated circuit modules are popular. These include the appropriate number of i-f stages, intercarrier sound takeoff, and video detector. Video amplifier ICs are also available. Sometimes the video section of a television receiver is a combination of integrated circuit and one or more discrete video stages.

At some point in the video amplifier (Fig. 8-1), composite signal is removed and applied to the sync and intersync separator. This can be an integrated circuit and is often referred to as a jungle IC.

First, the composite sync signal is removed from the combined sync-video signal that comes from the video amplifier. Then, the composite sync signal is divided into separate horizontal- and vertical-sync signals for application to the separate horizontal- and vertical-sweep generators. In present-day models, integrated horizontal- and vertical-sweep systems, although feasible, are not yet common. Integrated circuits are more common in vertical-sweep systems than in horizontal-sweep systems. Automatic gain control and automatic frequency-control systems are well adapted to integrated-circuit applications.

Although the integrated circuits used in television receivers are mostly special application devices, many of them are unusually ver-

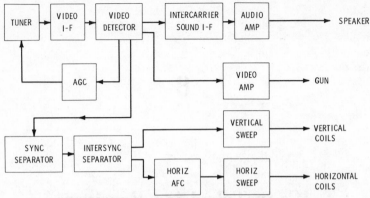

Fig. 8-1. Functional plan of a black and white television receiver.

satile. For example, video amplifier and i-f ICs can be used in other types of receiving systems, and various phase detection and comparison ICs can perform assignments in other electronic systems.

The function diagram of a color receiver (Fig. 8-2) depicts a similar system, plus the addition of two or three activities that are required for color demodulation and processing. Again, integrated-circuit tuners

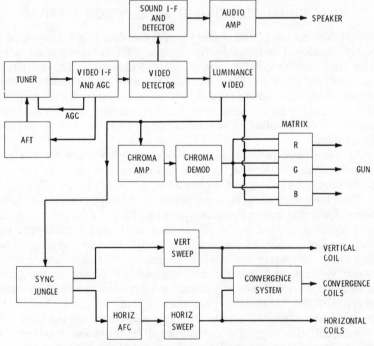

Fig. 8-2. Functional plan of a color television receiver.

are not common, although the associated automatic gain-control circuits and the automatic fine-tuning function are handled by integrated circuits. The former usually becomes a part of the integrated-circuit video i-f system.

In numerous color receivers, an integrated circuit is used for part of the i-f system, while the detector and the latter stages of video amplification are discrete-component circuits. The integrated circuit sound i-f amplifier and i-f detector is popular. It can be followed by a completely integrated audio amplifier or by a combination of integrated-circuit preamplifier followed by a discrete component audio-output stage, which drives the speaker.

From the video detector or a succeeding video stage, information must be channeled in several directions. The luminance video amplifier, which can be an integrated circuit or a hybrid combination, builds up the level of the video picture that determines the brightness variation and the detail in the reproduced color signal.

Both a component of composite television signal and the 3.58-MHz color-sync component required by the chrominance demodulation system must be applied to the usual integrated-circuit jungle system, which forms the composite sync and intersync component. There must also be a path for the chrominance signal itself, which must be segregated into demodulated components. These components eventually combine with the luminance signal to form the tristimulus red, green, and blue primary color signals that are applied to the separate guns of the color picture tube. Integrated circuits are used widely in the chrominance demodulation and color-processing circuits.

An integrated circuit is often used in the horizontal automatic-frequency-control system of the receiver. However, for the most part, the vertical-deflection, horizontal-deflection, and convergence circuits of present-day color receivers use discrete-component stages. However, complete integrated-circuit deflection systems are certain to become increasingly popular. IC regulators can also be found in the power systems of television receivers.

In the material that follows, individual specialized television ICs are discussed. Considerable details are also given about how they are used and fitted into the complete television receiver package.

ICs IN TELEVISION I-F SYSTEMS

Integrated circuits are common in both the video and sound i-f systems of television receivers, because they greatly simplify and stabilize these circuits. Only a minimum number of discrete external components are needed, mainly bandpass transformers and tuned circuits. An example of a 20-pin integrated circuit that includes both the video and audio i-f circuitry is the RCA CA3068 IC (Fig. 8-3).

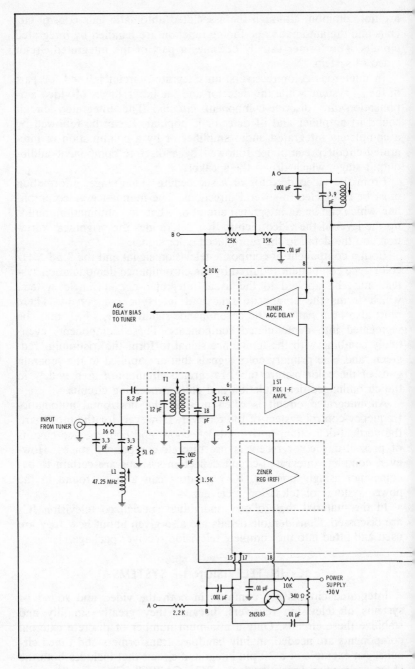

Fig. 8-3. RCA CA3068 integrated

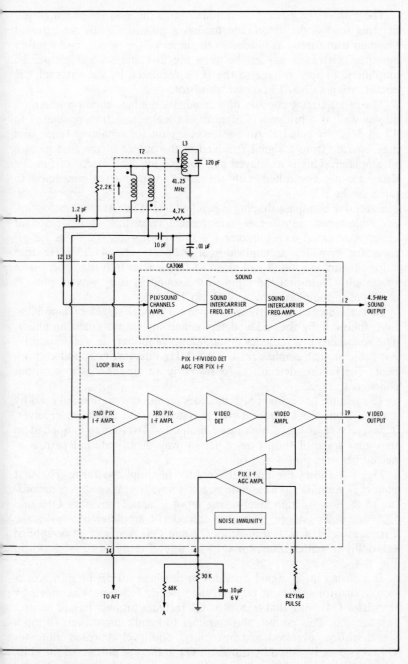

circuit i-f system.

The gain of the picture i-f amplifier is 75 dB, and the video amplifier that follows the video detector has a gain of 12 dB. An external resonant transformer is needed at the input of the device, and another bandpass system is needed between the first and second picture i-f amplifiers. Supply voltage to the IC is regulated by the external VR circuit, using a 2N5813 bipolar transistor.

The input circuit consists of a bandpass double-tuned transformer, T1, as well as a bridged-T adjacent-channel sound trap resonated to 47.25 MHz by coil L1. An agc delay circuit for obtaining tuner bias may operate from a signal removed at the output of the first picture i-f amplifier. This is a delayed agc component. The extent of the agc delay can be controlled with the 25K potentiometer connected to pin 8 of the IC.

Extensive bandpass shaping is handled by the interstage transformer, a double-tuned transformer arrangement consisting of the resonant circuit associated with inductor L2 and transformer T2. The 1.2-pF capacitor provides a bandwidth adjustment. The 41.25-MHz trap helps channel the sound and video i-f frequencies into the appropriate i-f systems. Note that the input to the sound i-f is by way of pin 12; the input to the picture by way of pin 13.

In the video channel there are two additional stages of amplification, followed by the video detector and an output video amplifier. The sound channel includes an additional amplifier, as well as a mixing detector that emphasizes the 4.5-MHz intercarrier sound component. The mixing detector is followed by an amplifier whose output connects to pin 2.

The picture i-f segment also includes an agc system that is keyed by a pulse from the horizontal-sweep system of the television receiver. This agc system has a high noise immunity. Its voltage is applied to the input stage of the i-f amplifier by way of an external path connected between pins 4 and 6.

Fig. 8-4 shows the internal circuitry in simplified form. The first picture i-f amplifier is a cascode circuit whose operation differs according to the low or high level of the input signal. Transistors Q19 and Q20 are buffer stages, while the cascode i-f amplifier comprises the circuit associated with transistors Q2 and Q3. Agc voltage is applied externally to buffer transistor Q19 by way of terminal 6, as shown in Fig. 8-4.

If a strong input signal provides as much as 40 dB of gain reduction, transistor Q3 is cut off, and the cascode function is assumed by transistor Q4. Note that transistor Q4 includes an unbypassed emitter resistor, R6. This permits the amplifier to handle high levels of input signal without overload and instability. The level at which transistor Q4 takes over is sensed by transistor Q1 in the agc portion of the chip. At this time, transistor Q21 is also activated, shutting off Q5 and

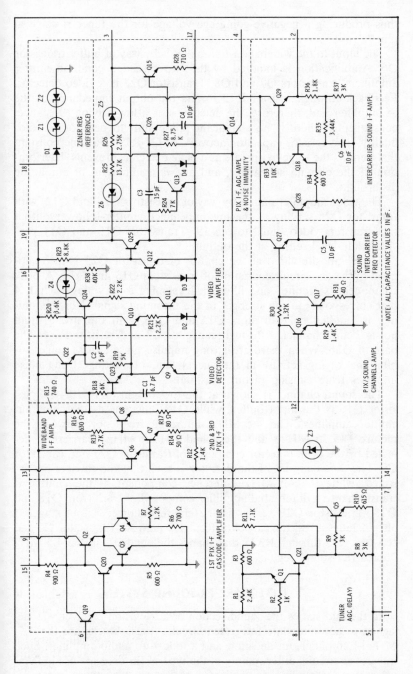

Fig. 8-4. Internal schematic of RCA CA3068 IC.

285

thus producing a negative-going agc voltage for the tuner rf stage at pin 7.

The input to the wideband i-f amplifier is by way of buffer transistor Q6. Amplification is handled by the two cascaded common-emitter amplifiers, transistors Q7 and Q8. Transistor Q22 is a video detector. Feedback through resistor R13 lowers the output impedance of the i-f amplifier and, therefore, the detector sees a low driving resistance; consequently, there is no significant phase shift to deteriorate color fidelity. The resistor-capacitor network of C2-R19 is a time constant that permits the stage to operate as a peak detector, prevents deterioration of the chroma subcarrier, and allows a minimum of amplitude distortion.

The video amplifier path is by way of transistors Q24 and Q12, with transistor Q25 serving as the low-impedance emitter-follower output that transfers video signal to pin 19. Transistors Q10 and Q11 provide white-level setting at the output.

The composite video at the emitter of transistor Q25 is also supplied to the agc circuit consisting of transistors Q13 and Q14. A peak component is rectified and filtered, developing a dc voltage at pin 4 that corresponds to the strength of the incoming signal. Noise immunity is handled by transistors Q15 and Q26, with the strong input noises being channeled through capacitor C3 and resistor R7 to the base of Q26. When there are strong impulse noises, these stages inhibit the keying activity in the agc system, preventing strong noise impulses from causing picture washout.

The sound channel is shown at the lower right of Fig. 8-4. Transistor Q16 is the input buffer, while Q17 functions as the common-emitter amplifier. The 4.5-MHz difference frequency between the picture (45.75 MHz) and the sound (41.25 MHz) carriers is detected by the peak detector composed of transistor Q27 and capacitor C5. The time constant of the detector is such that the carriers are rejected, while the 4.5-MHz difference component at the emitter is transferred to a differential amplifier composed of transistors Q18 and Q28. Transistor Q29 serves as the emitter-follower output stage. It also includes a resistor-capacitor feedback network (capacitor C6 and resistors R35 through R37) that shapes the bandpass in the 4.5-MHz intercarrier-sound region.

TV INTERCARRIER-SOUND SYSTEM

In the mid-sixties the sound section of a television receiver was designed to accommodate integrated circuits. Basic blocks are the 4.5-MHz i-f amplifier, fm detector, and a follow-up audio amplifier. Such integrated circuits differ mainly in the type of fm detector used. Typical systems, including discriminator and quadrature types, were de-

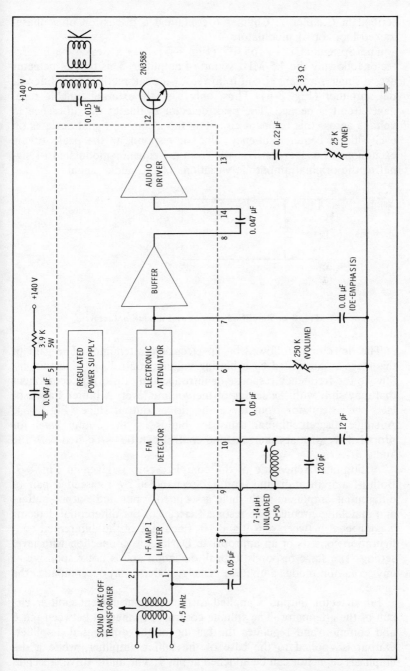

Fig. 8-5. An intercarrier sound system using a Sprague ULN-2165 IC.

287

scribed in Chapter 7. Chapter 6 contained a discussion of a phase-locked-loop fm demodulator.

The Sprague ULN-2165 IC (Fig. 8-5) uses a peak-type fm detector following the 4.5-MHz sound i-f amplifier. This type of detector uses a single resonant circuit followed by a peak rectifier and differential amplifier (Fig. 8-6). Thus, only a single external tunable resonant circuit is needed. The peak detector evaluates the difference in voltage across this resonant circuit as the incoming signal deviates on each side of center frequency. Since it responds to the peaks of the excursions, a conversion is made from a frequency-modulated i-f signal to the original amplitude variations of the audio signal.

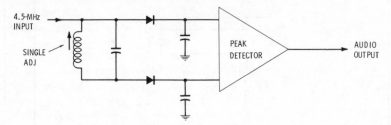

Fig. 8-6. Simplified plan of an fm peak detector.

The detector is followed by an electronic attenuator. The gain of this stage is controlled by a biasing volume control that does not influence the frequency response or introduce any distortion component that may shift with the setting of the volume level. A buffer stage isolates the attenuator from the audio driver output stage (Fig. 8-5). A single discrete bipolar transistor builds up the output level for proper drive of a speaker. A tone control can be associated with the audio driver.

A simplified schematic of the complete chip is given in Fig. 8-7. Both i-f amplification and limiting are handled by a cascaded pair of differential amplifiers with their associated input and output buffers and matching circuits. The resonant circuit of the differential detector is connected between pins 9 and 10. Each side of the differential configuration consists of an appropriate Darlington connection with level setting. The stage responds to peak deviation and is not responsive to any common-mode amplitude changes that may accompany the signal.

Fm detector output is applied to the constant-current emitter circuit of the attenuator. The volume control is connected between pin 6 and common and regulates the biasing of the differential amplifier. Output is applied to the base of the buffer amplifier, while a de-emphasis capacitor can be attached at pin 7. The audio driver receives its excitation when a capacitive coupling is established between pins 8

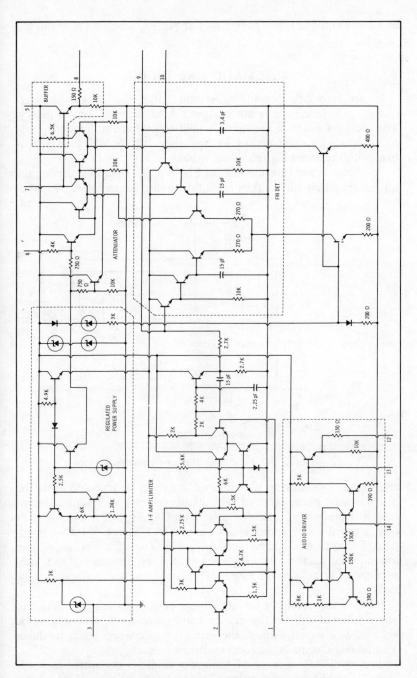

Fig. 8-7. Internal circuit of the Sprague ULN-2165 IC.

289

and 14. Amplified audio is removed at pin 12, which connects to the emitter.

AUTOMATIC FINE TUNING

In an automatic fine-tuning system (aft), the intermediate-frequency picture carrier is automatically held precisely on frequency. The local oscillator of the tuner is held exactly at the frequency that will produce an exact picture i-f frequency (45.75 MHz) when it is beat with the incoming television signal.

As shown in the function plan of Fig. 8-8, the signal for driving the aft integrated circuit is taken from the picture i-f amplifier. An adja-

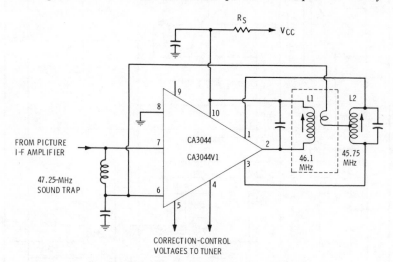

Fig. 8-8. Functional plan of aft system.

cent-channel sound trap is positioned at the input. Included in the IC is a balanced phase detector with its phase-sensitive resonant circuit tuned to the picture carrier frequency of 45.75 MHz. The primary side is tuned to the approximate center (46.1 MHz) of the i-f bandpass. Dc correction-control voltages are developed at the output and are supplied back to the frequency-controlling elements of the local oscillator in the vhf-uhf tuner. These are usually voltage-variable capacitor diodes.

Because the output of the aft IC is a dc voltage that is sensitive to any attempted shift in the picture carrier frequency, this shifting dc voltage, as it passes through the variable-capacitance diode, produces a correcting change in the local-oscillator frequency.

The input stage is a 45-MHz limiter amplifier connected in a differential-amplifier configuration (Fig. 8-9). The primary of the phase-

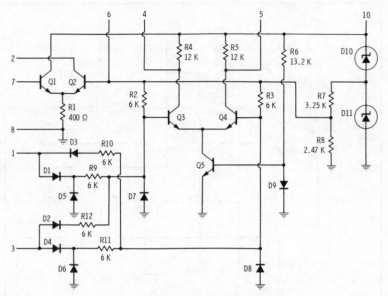

Fig. 8-9. An RCA CA3044 aft IC.

detector transformer connects in the collector circuit of transistor Q2 at pin 2 (Fig. 8-8). The ends of the balanced 45.75-MHz secondary connect across pins 1 and 3 of the diode balanced detector consisting of diodes D1 through D4. D5 and D6 are balanced diodes compensating for the parasitic diode capacitances that are present between the cathodes of diodes D2 and D3 and the substrate.

Diodes D7 and D8 function as capacitors that filter the output of the detector in conjunction with resistors R9 through R12. Thus, a dc component is made available as a function of the relationship between the carrier component and the resonant frequency of the transformer secondary. This dc component is applied to the bases of the constant current dc amplifier, which is composed of transistors Q3 and Q4, plus constant-current-source transistor Q5, and its temperature-compensating diode D9. The dc correction voltages for the local oscillator can be removed at pins 4 and 5. A built-in voltage regulator uses zener diodes D10 and D11.

CHROMA PROCESSING AND DEMODULATION

Two or three integrated circuits can handle the operations that must occur between the chrominance takeoff position in the luminance video amplifier and the three separate red, green, and blue inputs to the preamplifier, which is ahead of the gun of the color picture tube. In a two-step process (Fig. 8-10), chroma processing and chroma color

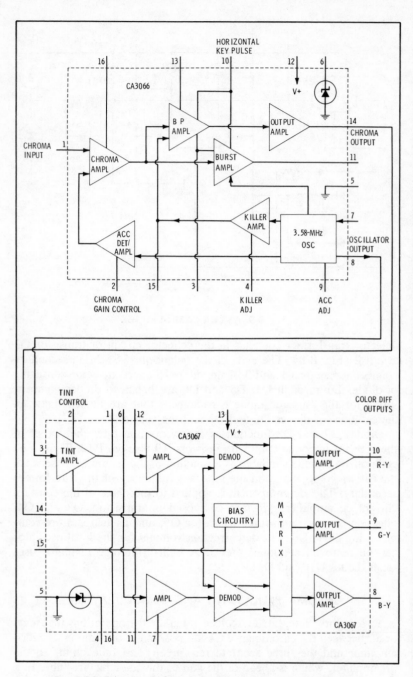

Fig. 8-10. Functional plan of chroma sections of a color receiver.

demodulator ICs are used. Examples are the RCA CA3066 and CA3067.

The chrominance signal to be applied to pin 1 is derived from an appropriate point in the luminance video amplifier of the color receiver. The luminance signal information has already been de-emphasized. The frequency-response range of the chroma amplifier and the bandpass amplifier is approximately 3.08 to 4.08 MHz. This is accomplished by connecting stagger-tuned coils to pins 13 and 16 (Fig. 8-11). Note the chroma input signal is capacitively coupled to pin 1.

Automatic chroma control (acc) is used to regulate the gain of the chroma amplifier in accordance with the strength of the incoming color-burst signal. As previously noted, the color synchronization information is transmitted as a brief 3.58-MHz burst, positioned at an appropriate point on the back porch of the horizontal-blanking pulse. The burst synchronizes the 3.58-MHz oscillator (Fig. 8-11). In Fig. 8-10 note that a line extends from the 3.58-MHz oscillator, which is under control of the incoming 3.58-MHz color-burst signal, to the acc detector and amplifier. A manual chroma-gain control is connected to pin 15 of the IC (Fig. 8-11).

The color-killer amplifier also responds to the presence of the color-burst signal. If the color-burst signal is not present, the killer circuit shuts off the bandpass amplifier and prevents stray signal from passing through the chrominance channel when the receiver is operated in its black and white mode.

An output from the chroma amplifier passes through the bandpass amplifier and through the output amplifier to the chroma input of the color demodulator (pin 14 of Figs. 8-10 and 8-11). The output of the chroma amplifier is also applied to the burst amplifier, which is adjusted to emphasize the 3.58-MHz frequency range. In so doing, the burst-frequency signal is amplified and can be used to control the 3.58-MHz oscillator of the signal processor. In fact, this oscillator now generates a continuous color-carrier component that needs to be reinserted in the demodulation of the chrominance signal.

A horizontal-keying pulse is applied to the burst amplifier by way of pin 10. Its function is to make the burst amplifier operative only during that short interval of time corresponding to the time interval of the burst frequency as it arrives on the back porch of the horizontal-blanking pulse.

In the color demodulation process, the hue of the reproduced color on the color picture-tube screen is the function of the angular relationship between the two chrominance signals, I and Q, which make up the incoming chroma signal. Color saturation is a function of their relative magnitudes. It is important, then, that the angle of the regenerated 3.58-MHz subcarrier, conveyed between pins 8 and 3, be

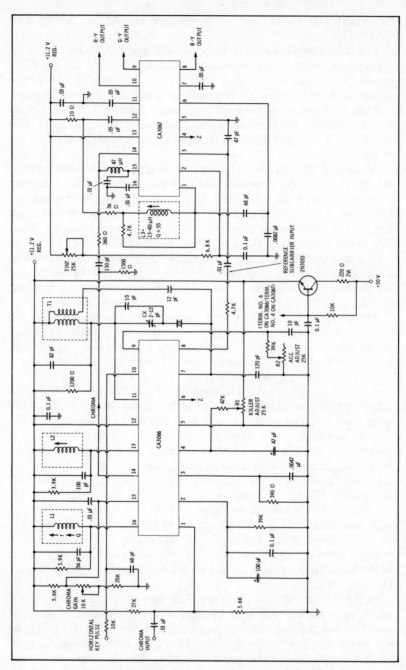

Fig. 8-11. Pin-out and external components diagram.

properly set to obtain a true representation of color hue and saturation. Adjustment is aided with the tint amplifier and associated tint control that is attached to pin 2 (Figs. 8-10 and 8-11).

The chroma signal is applied directly to the twin demodulators (doubly balanced demodulator) through pin 14. The regenerated sub-carrier is applied at pins 6 and 12, which obtain their signal from pin 1 of the tint amplifier (Fig. 8-11). The outputs of the demodulator are the original $R - Y$ and $B - Y$ chrominance signals. The $G - Y$ component is formed by matrixing $R - Y$ and $B - Y$.

The three chrominance outputs at pins 8, 9, and 10 are now combined with the luminous (Y) signal. In the combining process, the original red, green, and blue (RGB) tristimulus color signals are formed and applied to the separate color guns of the color picture tube (Fig. 8-12).

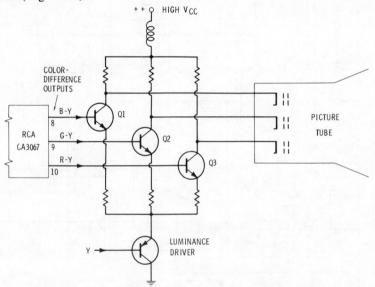

Fig. 8-12. Combining color difference and luminance signals to form red, green, and blue signals for the picture tube.

The internal circuitry of the RCA CA3067 is shown in Fig. 8-13. Three operational functions are performed: amplification, balanced chroma demodulation, and tint (phase) control. The reference sub-carrier is applied to pin 3; the chroma signal to pin 14. The three color-difference signals are removed at pins 8, 9, and 10.

The color subcarrier is differentially amplified by transistors Q2 and Q3. Resistor-capacitor combination R1-C1 and stray capacitances produce a phase delay of approximately 45° in the collector of transistor Q2. Thus, two reference subcarrier phases are available, the

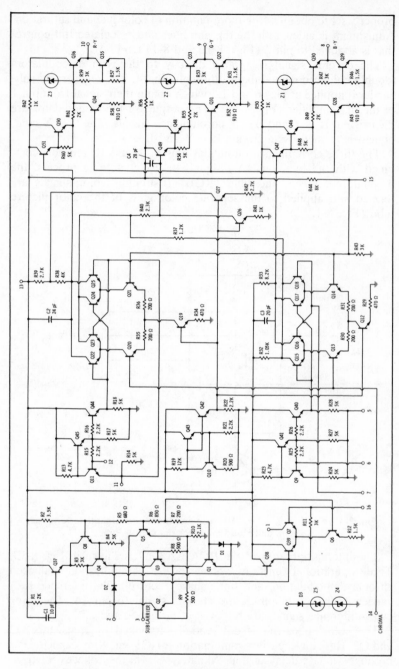

Fig. 8-13. Internal circuitry of RCA CA3067 color-demodulator IC.

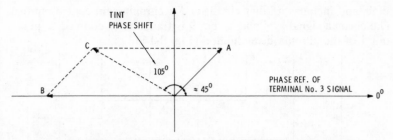

Fig. 8-14. Phase setting of the color demodulator reinserted carrier.

burst phase at Q3 and delayed phase at Q2. Eventually the two signals are recombined in the collector of transistor Q4. However, relative amplitude and phase determine the eventual phase of the color subcarrier that is applied to the base of transistor Q38, lower left. This angle is set at approximately 105° (Fig. 8-14). The actual angle can be set with the tint control, which adjusts the applied current by way of pin 2 (Figs. 8-11 and 8-13). Based on transmission standards and improved color phosphors, this is now a preferred reference for demodulation parameters.

Transistors Q37 through Q39 function as a limiter-amplifier, removing any amplitude modulation from the reference subcarrier signal. Output is removed at terminal 1 and, through an external coupling circuit, is applied to the demodulator by way of pins 6 and 12 (Fig. 8-11). The associated RLC circuit between terminals 6 and 12 provides a phase separation of 76°. As a result, there are two demodulation angles of 0° and 104° (180° − 76°) for the B − Y and R − Y demodulators.

The chroma signal is applied to a pair of chroma amplifiers and switchers; the input transistors to these two channels are Q13 and Q20. Transistors Q22 through Q25 serve as the R − Y demodulator; transistors Q15 through Q18 serve as the G − Y demodulator. Each channel provides both negative and positive output color-difference signals. The matrixing circuit combines the negative outputs of the R − Y and B − Y demodulators to develop the G − Y color-difference signals in accordance with transmission standards.

The three color-difference signals are applied to individual output amplifiers, which employ feedback to hold the output impedance low during both the positive and negative peak-signal swings. Drive capability is sufficient for exciting the high-voltage output transistors of present-day color receivers.

Sprague makes available a set of three integrated circuits—chroma amplifier, subcarrier regenerator, and chroma demodulator. Function diagrams for each are given in Fig. 8-15; a completely wired color segment is shown in Fig. 8-16. The chroma amplifier is a two-stage affair and includes a killer circuit and a chroma-gain control system. The chroma signal available at pin 9 is transformer coupled to pins 3 and 4 of the chroma demodulator (Fig. 8-16).

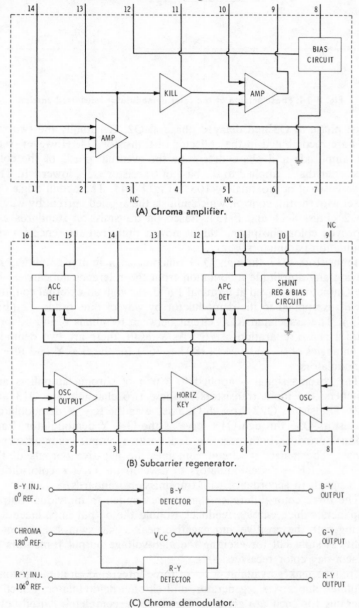

(A) Chroma amplifier.

(B) Subcarrier regenerator.

(C) Chroma demodulator.

Fig. 8-15. Block plan of Sprague chrominance ICs.

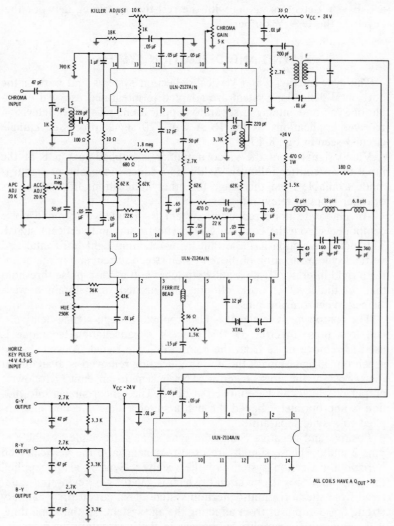

Fig. 8-16. Interconnection diagram of Sprague ICs.

The chroma regeneration system uses a phase-locked oscillator. It is controlled by the keyed automatic phase-control detector, which responds to the incoming color-burst signal when activated by the horizontal keying pulse during the appropriate horizontal back-porch interval. The oscillator output circuit and an associated hue-control facility set the angle of the regenerated subcarrier prior to its application to the color demodulator. An acc detector is also included in the subcarrier regeneration IC. The color demodulator includes the $R - Y$

and B − Y detectors, along with the resistive matrixing network that develops the G − Y output.

SYNC AND DEFLECTION SYSTEMS

The sync-intersync and vertical-deflection systems and part of the horizontal-deflection system are adapted to integrated use. An example of sync and intersync separator application, plus other television receiver applications, is the RCA CA3120 signal processor (jungle circuit) seen in Fig. 8-17.

Video signal from the video amplifier is applied to pin 8 at the input of an emitter follower. A low-impedance video output signal is made available from this stage for other applications in tv receivers or other types of television gear.

An elaborate noise-reduction system is included to remove noise components from the composite-sync (horizontal and vertical) signal. This noise reduction is important in maintaining tight horizontal and vertical synchronization of the television sweep system and in maintaining a rigid interlace. Both a pulse stretcher and a noise-pulse threshold that set the level of noise clipping are needed to maintain a well-shaped synchronizing pulse.

The composite-sync signal is also applied through a time-delay circuit to a noise inverter. In this process, a cancellation technique is used to remove noise from the composite-sync signal. A well-shaped synchronizing pulse with the noise reduced or removed is available at the output of the noise inverter. This is sent to an emitter follower, output of which can be removed at pin 5. The synchronizing information is appropriately shaped through a suitable time constant and applied to a sync separator.

Positive and negative composite-sync signals are made available at pins 2 and 3. These signals are used to synchronize the vertical- and horizontal-deflection systems of the receiver. Sync is also reapplied through pin 1 to a strobe circuit, which is gated by a keying pulse derived from the horizontal-deflection system. This circuit prevents very strong noise impulses from affecting the agc system. At the same time, it does permit the amplitude changes in the horizontal-sync information to determine the level of the agc bias. Agc bias, of course, sets the gain of the i-f and rf systems of the receiver, in accordance with the strength of an incoming television signal. Positive or negative agc bias is made available in accordance with the type of tuner used by the receiver, bipolar transistor, or MOSFET.

Motorola makes available integrated circuits designed specifically for the vertical- and horizontal-deflection systems of a television receiver (Fig. 8-18). The vertical-deflection system is virtually complete and provides direct drive for the vertical-deflection (scan) coils.

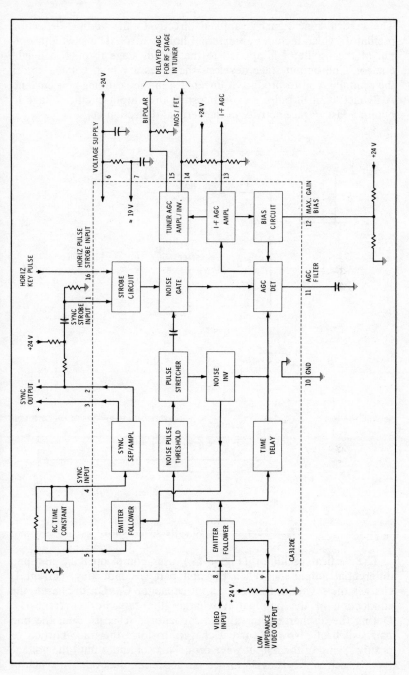

Fig. 8-17. RCA CA3120 IC used as a sync-signal processor.

The vertical-sync signal is applied directly to the vertical-deflection oscillator, which includes an external hold control. The sawtooth output of the oscillator is applied to the output stage through a height control. The output stage develops the necessary high voltage across the scanning coil to produce a linear rise in the scanning-coil current. A flyback control pulse aids in boosting the output-circuit voltage to obtain a fast retrace activity in the vertical-deflection coil.

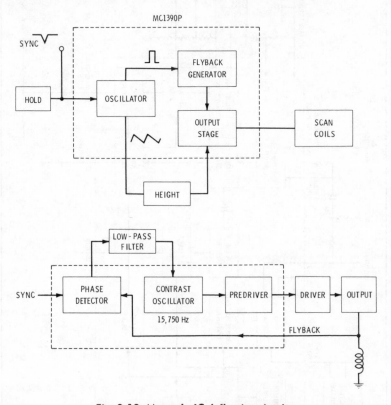

Fig. 8-18. Motorola IC deflection circuits.

The vertical oscillator (Fig. 8-19) uses complementary npn-pnp differential amplifiers. When the npn pair is conducting, current I_1 charges the sawtooth-forming output capacitor C_O. On discharge, the conduction of the pnp pair discharges the capacitor (current I_2). During the discharge interval, the capacitor discharges until the pnp pair switch off. Now I_1 current charges to form the trace portion or positive ramp of the output waveform. This continues until the voltage level is such that I_2 is again turned on, repeating the cycle. In synchronized operation, it is the incoming negative vertical-sync pulse that

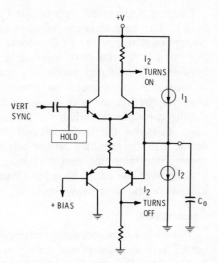

Fig. 8-19. IC vertical oscillator.

triggers the oscillator into operation at repeated intervals, thus establishing the necessary synchronized frequency of operation.

The horizontal-sync system is an automatic frequency-control arrangement using a phase detector to compare the incoming horizontal-sync pulse with a flyback component derived in the output stage. This phase detector develops a dc voltage component that is transferred through a low-pass filter to the control oscillator. Any change in the relative phase between the two components results in a dc voltage that makes the appropriate correction in the horizontal-oscillator frequency.

A predriver stage builds up the level of the signal from the horizontal oscillator. The driver and sweep power-output stages are external to the integrated circuit.

The previous discussions show the wide applications of integrated circuits to television receivers and other television gear. It is anticipated that 80 to 90 percent of the functional circuits in a television receiver will soon be in the form of integrated circuits.

PROJECT 3: PLL BROADCAST DETECTOR

General

A variety of electronic operations can be performed by a phase-locked loop. The PLL can be used to demodulate a-m, fm, and various other forms of modulated signals. In this project, the PLL is used as an a-m demodulator for the standard broadcast band. In fact, the circuit demonstrates how a phase-locked loop can be used to sort incoming signals without the need for any type of resonant input circuit.

Chapter 6 discussed a particular PLL, the Signetics NE561B (Fig. 6-16). The demodulation of an a-m signal is effective because the control oscillator of the PLL is locked to the frequency of the incoming carrier. As you have learned, a direct conversion is then made between a modulated radio-frequency signal and the demodulated audio signal. It is possible to tune over the broadcast band simply by changing the frequency of the vco with an external variable capacitor. As the capacitor is varied, the frequency of the internal oscillator is made to match the frequencies of the various broadcast carriers receivable in the area. By proper selection of variable capacitance and appropriate minimum and maximum limits, the device can be made to tune over the entire broadcast band or a segment of it, as well as long-wave and short-wave spectra.

Construction

The PLL a-m broadcast detector can be mounted on the same vector board as the audio power amplifier constructed in Project 1 (Fig. 8-20). Actually, very few components are needed and the results are exceptionally good (Fig. 8-21). Both a large and a small variable capacitor are recommended, with the small one being used for bandspread tuning. No tuning coil is required. Construct the circuit.

Fig. 8-20. Adding a PLL detector ahead of the audio amplifier.

Operation

The vco oscillator frequency is controlled by the capacitors connected between pins 2 and 3. The fixed 220-pF capacitor (C6) sets the highest frequency that can be received. This frequency is reached

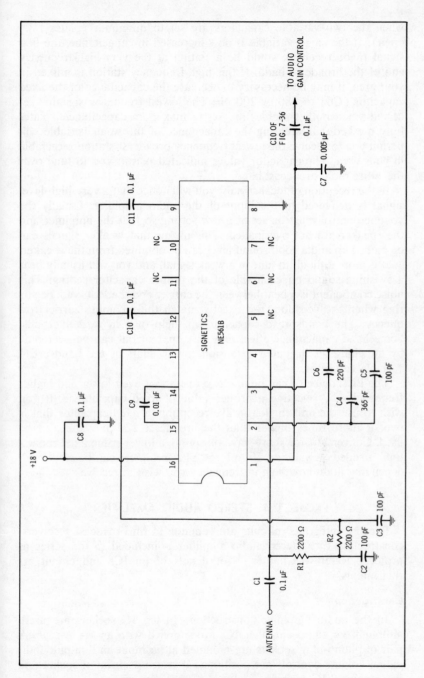

Fig. 8-21. PLL a-m broadcast detector.

when the two variable capacitors are set to minimum (plates fully open). If the small variable is now increased in capacitance, the first signal to be received would be a station at the very-high-frequency end of the broadcast band. If this high-frequency station is missed in your area, it may be necessary to decrease the capacitance of the fixed capacitor (C6) to 180 or 200 pF. The lowest frequency signal is received when both variables are set to maximum capacitance (plates fully meshed). Decreasing the capacitance of the small variable will permit you to receive the lowest frequency broadcast station receivable in your area. The capacitor values indicated permit you to tune over the entire radio broadcast band.

In the reception of local signals, you will notice that a very-high-level signal is delivered to the input of the audio amplifier. Usually the volume control must be set at a low setting so that the amplifier and the speaker are not overloaded. This means that weaker signals can be picked up and a good sound level can be obtained from the speaker.

It is more difficult to tune in a weak signal, and you will usually hear a swishing sound on each side of the proper capacitor setting. This noise component is a beat between the carrier and the PLL vco, occurring when the vco does not exactly match the incoming carrier frequency. The bandspread capacitor will help tune in weaker signals. Some hand capacitance effect may be apparent but can be overcome by attaching an insulated extension to the shaft of the bandspread variable capacitor.

The PLL detector can be made to operate at even lower and higher frequencies by selecting a proper value of fixed capacitance. If you wish to use the unit for long-wave reception, a fixed capacitor that is two or more times higher than the broadcast 220-pF value can be used. Conversely, for short-wave reception, a lower-value fixed capacitor is needed. A value of 100 pF, 50 pF, and lower can be employed, depending upon how high in frequency you wish to receive.

PROJECT 4: STEREO AUDIO AMPLIFIER

Because integrated circuits are common in fm broadcast receivers, construction of a stereo audio amplifier is included in the series of projects. Refer to Chapter 7 for details of fm ICs and circuit arrangements.

Construction

In the audio amplifier construction, 14-pin IC sockets are used, although the audio amplifier ICs are supplied with an 8-pin case. A pair of phono-plug sockets are mounted at an input and output. Individual volume controls are positioned between each input socket and its associated IC socket.

Few external components are needed (Fig. 8-22) to construct this complete audio amplifier, which delivers a half-watt of audio power output with an input signal of only 3 millivolts (rms).

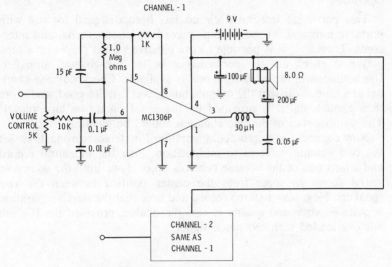

Fig. 8-22. Schematic of stereo audio amplifier.

Table 8-1. Parts List for Projects 3 and 4

1	IC-12 (from Project 1)
1	Signetics NE561B IC
2	Motorola MC1306P IC
3	14-pin in-line IC sockets
1	365-pF variable capacitor
1	100-pF variable capacitor
2	24-μH rfc (Miller 4626)
1	4½″ × 8½″ vector board
4	Phono sockets
2	5K potentiometers
2	8-ohm speakers
2	9-volt batteries
6	Binding posts
2	1K, ½-watt resistors
2	22K, ½-watt resistors
2	10K, ½-watt resistors
2	1-meg, ½-watt resistors
2	15-pF disc capacitors
2	100-pF disc capacitors
1	220-pF disc capacitor
1	0.005-μF disc capacitor
3	0.01-μF disc capacitors
6	0.1-μF disc capacitors
1	100-μF 25-volt electrolytic capacitor
2	200-μF 25-volt electrolytic capacitors

Construct the circuit, keeping the two channels isolated from each other and suitably separating the input and output of each channel.

Operation

This particular integrated circuit has been designed for use with portable a-m and fm radios, tape recorders, phonographs, and intercoms. Therefore, it is possible to use either a phono player or a tape player to check out the performance of the two-channel amplifier. Use a monaural record first, if one is available. Check out one channel at a time. Note that the output sound level is quite good, and more than enough signal is provided by the usual phono or tape player. The performance of the two channels should be identical.

Now connect the stereo audio amplifier for stereo reproduction. Set the two volume controls to midposition. Play the monaural record and adjust one of the volume controls up or down until the source of sound seems to come from the center position between the two speakers. Now play a stereo record and note that the stereo separation is quite good. Sound quality is also reasonable, provided the ICs are not overloaded with too much signal.

9

Industrial and Instrumentation
Systems

The integrated circuit is a compact, stable and versatile device. These qualities are particularly valuable in the operational amplifier, which often provides the critical control circuitry of industrial electronic systems and instrumentation.

Modern electrical instruments employ a variety of linear and digital ICs. The latter devices are particularly helpful in the counting and display systems of various electronic meters and generators.

GENERAL

There are three basic blocks associated with industrial electronic-control systems (Fig. 9-1A). The first one is a sensor that responds to a particular parameter, which can be temperature, humidity, light, flow, pressure, vibration, velocity, etc. Sensors can be discrete devices or perhaps integrated-discrete hybrid combinations. Increasingly, the control and/or programmed sections have become integrated circuits.

The second basic block is the control section, for which both generalized and specialized IC devices have been developed. The main function of the control section is to respond to a sensor signal in a programmed manner and then activate appropriate power circuitry.

Power devices comprise the third basic block associated with industrial electronic-control systems. They also come in a variety of forms, from a simple switch to complex feedback power systems. Typical power devices are relays and solenoids, indicators, motors and generators, heating elements, cooling elements, valves, flow regulators, fans, blowers, level setters, etc. Under the control of a very weak sig-

(A) Three basic blocks.   (B) Instrumentation package.

Fig. 9-1. Basic control and instrumentation functions.

nal from a sensor, a low-powered integrated-circuit control system can be made to switch modest levels of power.

An instrumentation package is a similar three-block affair (Fig. 9-1B). Many modern electronic measuring instruments and generators are built almost completely around integrated circuits. Some sort of probe is used to take a sampling of whatever electrical/electronic parameter is to be measured. The control section evaluates and compares this sample and displays an electrical signal on a calibrated indicator, or provides a digital readout.

PROGRAMMABLE POWER SWITCH/AMPLIFIER

The RCA CA3094 IC (Fig. 9-2) is a combination power switch and amplifier that can deliver an average output of 3 watts and a peak output of 10 watts. This amount of power can be obtained with

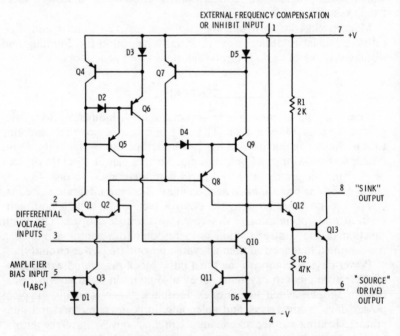

Fig. 9-2. RCA CA3094 programmable power switch/amplifier.

input signals measured in microamperes and millivolts. When quiescent, this monolithic IC dissipates only a few microwatts.

The IC consists of an output amplifier—transistors Q12 and Q13—and is preceded by an operational transconductance amplifier (OTA). It is so called because the output signal is in the form of a current that is proportional to the voltage difference at its input or:

$$g_m = \frac{\Delta i_{out}}{\Delta e_{in}}$$

The device has the usual differential input but, in addition, it has an amplifier bias input, pin 5, which can be used for either linear gain control or a position to which a gate pulse or voltage can be applied. Pin 2 is the inverted input; pin 3, the noninverting input. The output of the OTA section can be observed at pin 1.

The bias current applied at pin 5 sets the emitter current of differential transistors Q1 and Q2. This bias determines the differential transconductance of the amplifier. The transistor pairs, Q8-Q9 and Q10-Q11, develop a current-controlled output and are designed in such manner that they have little voltage sensitivity. Transistors Q12 and Q13 serve as the Darlington-connected output and provide access to either the collector (sink) or emitter (source) terminals, depending upon the output characteristics desired.

THE CA3094 AT WORK

A basic industrial and instrumentation IC circuit is given in Fig. 9-3. Output voltage is a function of the degree of imbalance present in the input bridge. The transducer bridge can be any one of a number of sensor types; the sensor element is often just one leg of a resistive bridge network. A centering or reference control is associated

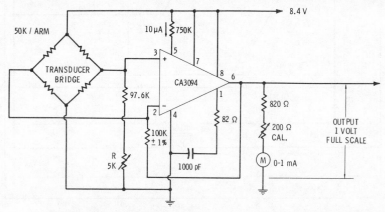

Fig. 9-3. The CA3094 IC used in a basic sensor-bridge circuit.

311

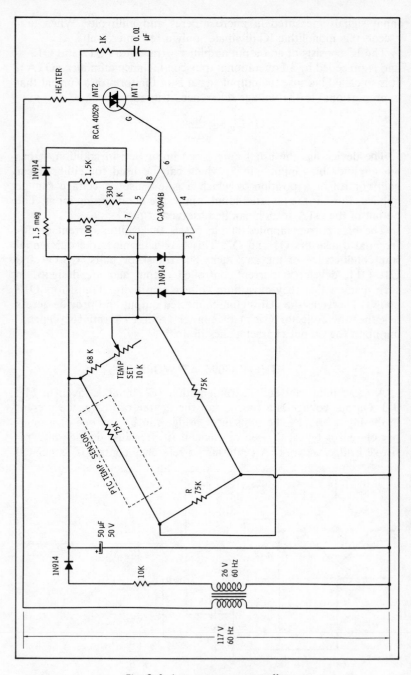

Fig. 9-4. A temperature controller.

with one leg of the bridge, while a second control, used for calibration, is associated with the output circuit, pin 6.

A practical temperature controller, using a resistive bridge network and including a "temperature-set" rheostat, is given in Fig. 9-4. The bridge has 75K resistive legs, one of which is a sensor resistor with a positive temperature coefficient.

The unit is operated directly from the ac line and incorporates a step-down transformer and single-wave rectifier. When the temperature drops, so does the resistance of the sensor resistor, making terminal 3 more positive than terminal 2. Under this condition, the output at pin 6 causes the triac to conduct. This, in turn, applies power to the heater. When the temperature of the device is brought up, the ohmic value of the sensor resistor increases and the temperature controller is shut down.

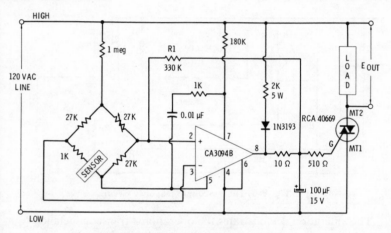

Fig. 9-5. A line-operated firing circuit.

If need be, a negative-temperature-coefficient sensor can be used simply by reversing the position of the sensor resistor and resistor R. In most of the control circuits, two diode rectifiers are placed across the input to prevent damage to the CA3094 in case of an excessive differential-input voltage.

The example of Fig. 9-5 shows a practical circuit that can be used when an ac sensor must be employed. Note that there is no step-down transformer or rectifier. Typical of many industrial-control circuits, the device is largely inoperative on one alternation of the ac line voltage. In this case, a negative ac line voltage shuts down the CA3094 because there would be no forward-bias current supplied to pin 5.

When the bridge is unbalanced in such a direction that terminal 2 is made more positive than terminal 3, the CA3094 is off when the ac line voltage swings positive. This causes the level at terminal 8 to go high. As a result, there is current through diode 1N3193 and the 1-μF capacitor is charged, driving the triac into conduction. When the negative swing of the ac line occurs, there is enough stored electrical energy to maintain triac conductance.

A bridge imbalance in the opposite direction causes CA3094 operation, and the current through the 1N3193 diode is conducted through the IC. The triac gate is not triggered.

A basic thermocouple-control circuit using a differential thermocouple is shown in Fig. 9-6. In this arrangement, the amount of heat sensed by the thermocouple develops a differential voltage between pins 3 and 2. In this manner, a conversion is made between a differential-input signal to a single-ended output signal. In fact, a differential-input voltage swing of ±26 millivolts results in an output current range of ±8.35 milliamperes.

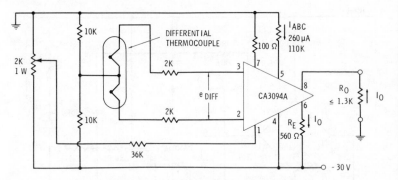

Fig. 9-6. Thermocouple response circuit.

This operation demonstrates the operational transconductance (output current change as a result of input voltage change) performance of the CA3094. The actual transconductance is controlled with the I_{ABC} dc current applied to pin 1 by way of the arm of the 2K potentiometer. In fact, a g_m value of 5000 micromhos results when this current has a value of 260 microamperes.

MULTIVIBRATOR AND VCO

In Fig. 9-7 the CA3094 is connected in an astable multivibrator circuit employed as a light flasher. For 25 percent of the time, the bulb is turned on with current drawn through terminal 8 (low state). A feedback path exists between terminal 8 and input terminals 2 and 3. The important frequency-determining time constant is set by resistor R and capacitor C. This time constant establishes the length of the high output at terminal 8 when no current passes through the flasher bulb. It is said that there is a 75 percent *off* time during which interval the

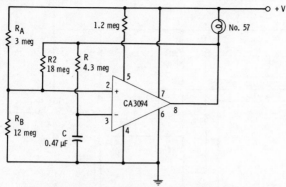

Fig. 9-7. The CA3094 IC used as a light flasher.

power drawn is of microwatt level. The current during the shorter 25 percent *on* period is in excess of 100 mA.

An advantage of the control influence of the bias current I_{ABC} is that it is able to control the frequency of operation of a multivibrator-type circuit by setting the transconductance of the IC (Fig. 9-8). Frequency of operation can be changed by varying the level of the current at terminal 5.

Current can also be changed by applying an input voltage through a suitable series resistor R. Now by changing the voltage, one changes the current and then the frequency of the relaxation oscillator, setting up a voltage-control oscillator circuit. Such an arrangement, of course,

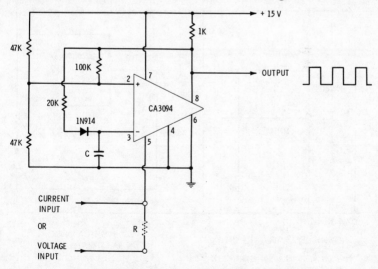

Fig. 9-8. Voltage- and current-controlled oscillator.

315

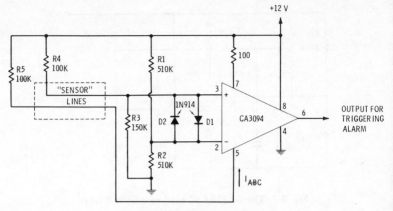

Fig. 9-9. Alarm circuit.

can be used in phase-detector and PLL applications. The resistive-feedback values and capacitor C determine the basic frequency of operation, which can then be varied with the changing current or voltage at terminal 5.

ALARM AND TIMER CIRCUITS

The alarm circuit of Fig. 9-9 operates with a high output at pin 6 in the no-alarm condition. The terminal-2 voltage is lower than the

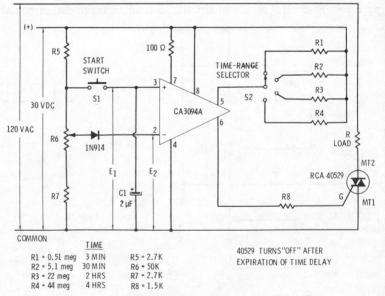

	TIME			
R1 = 0.51 meg	3 MIN	R5 = 2.7K		
R2 = 5.1 meg	30 MIN	R6 = 50K		
R3 = 22 meg	2 HRS	R7 = 2.7K		
R4 = 44 meg	4 HRS	R8 = 1.5K		

40529 TURNS "OFF" AFTER
EXPIRATION OF TIME DELAY

Fig. 9-10. Timing circuit.

potential at terminal 3, and the I_{ABC} current is sufficient to keep the output voltage high. When any one sensor line is open, is shorted to ground, or is shorted between lines, the output switches to low, and output current activates an alarm system. Any one of the three alarm conditions lowers the difference of potential between terminals 3 and 2.

The timing circuit of Fig. 9-10 is interesting because of its ability to set up very long timing intervals of as much as 4 hours. This is ac-

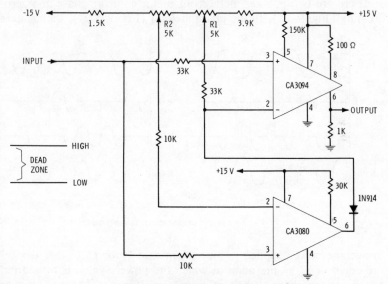

Fig. 9-11. Dual-limit detector.

complished by discharging input capacitor C1 through the IC by way of terminal 3. The level of the discharge current can be precisely controlled by the I_{ABC} current into terminal 5. Note that the time-range selector switch is connected to pin 5.

In placing the circuit in operation, the START switch is depressed, charging capacitor C1. When it is released, the capacitor begins its long discharge interval. The discharge continues until voltage E_1 becomes less than voltage E_2. Terminal 2 now draws current, reversing the polarity of the output voltage at terminal 6. In so doing, the output thyristor is activated. The 1N914 diode limits the maximum differential-input voltage.

The circuit of Fig. 9-11 is unusual in that it provides a 12-volt output whenever an applied input signal exceeds either a reference high limit or a reference low limit, as established by potentiometer R1 (high-level) and potentiometer R2 (low-level). Whenever the high limit is exceeded, the potential at terminal 3 of CA3094 increases relative to that at pin 2. The terminal-6 voltage then becomes a high 12 volts. A similar operation occurs when the input drops below the low limit, with the exception that the activity occurs through the CA3080 low-limit detector, which, in turn, changes over the CA3094 to the high-output condition.

MOTOR CONTROL WITH ICs

There are two popular methods of motor control: phase and zero-crossing (Fig. 9-12). In the phase method (Fig. 9-12A), the changes

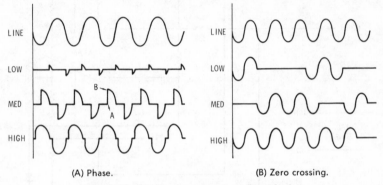

(A) Phase.   (B) Zero crossing.

Fig. 9-12. Two basic motor-control methods.

in average power applied to the system load (motor or other device) are determined by the point at which the power switch is turned on, relative to the phase of the load waveform. If power is turned on near the end of each alternation of the load waveform, only a very small power is delivered into the load. Maximum power is delivered when power is switched on for almost the duration of each alternation.

Two disadvantages of the phase-control method are its tendency to generate radio-frequency interference and the elaborate and costly filter system needed to take out such interference. In the phase method, there is an abrupt change in power that occurs instantaneously when the load current very quickly rises from zero to a particular operating level. This abrupt rise in the waveform between points A and B generates high-frequency harmonics and other spurious signal components.

The zero-crossing method (Fig. 9-12B), which is popular in integrated circuits, reduces the radio-frequency interference because the application of the power is made to be coincident with the passing of the load waveform through its zero axis. In this method, as shown in the waveforms, power is switched on for complete cycles of the load waveform. Little power is delivered to the load by switching on power for only one cycle of a sequence of load waveforms. Maximum power results when power is continuously held on, as in the last waveform.

An integrated circuit that can be used in such a zero-crossing control system is Fairchild's *Trigac* (Fig. 9-13). The functional sections are the sensor, the *Trigac,* and the thyristor power switch. The functional block diagram of the *Trigac* is also given.

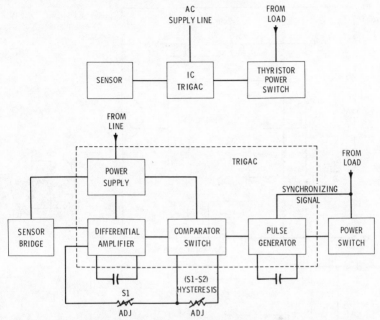

Fig. 9-13. Functional plan of Fairchild *Trigac* system.

The *Trigac* operates with a 10-mW sensor. It is able to switch load circuits up to a power of 5 kW. Power range can be extended up to 200 amperes with the addition of an appropriate external pulse transformer (Fig. 9-14).

The input circuit is planned for operation with a 10K sensor with supply voltage set to limit power to 10 mW. The amplifier is the standard IC differential amplifier, using a common-emitter current source to obtain excellent common-mode signal rejection and high operating stability under temperature change. Amplifier output is supplied to the comparator switch, the function of which is to set the time intervals for the application of power to the load. The switch output applies power to a pulse generator.

Two adjustments are included to set the power-on voltage level and the width of the hysteresis loop.

The purpose of the pulse generator is to form a pulse of short duration that coincides in time with the passing of the load waveform through its zero axis. The precise timing is accomplished by a synchronizing component that is fed back from the load to the pulse generator. Feedback pulse is feasible because at the very moment the load waveform is passing through its zero axis, as applied to the *Trigac* switch, there is a momentary loss of holding current and the resultant high resistance generates a synchronizing pulse.

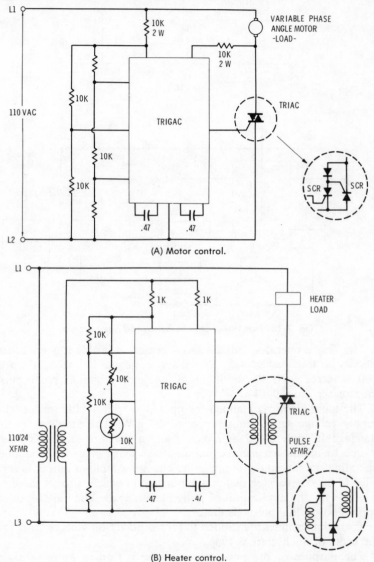

(A) Motor control.

(B) Heater control.

Fig. 9-14. Basic *Trigac* motor- and temperature-control circuits.

This component activates the pulse generator, which, in turn, supplies the necessary turn-on pulse to the gate of the thyristor. The thyristor, in turn, resets to the *on* position for the next half-cycle of the load waveform.

The use of this synchronizing pulse permits the circuit activities to follow any drift in phase of the load waveform and permits operation with the usual inductive type of load. Note the simple circuitry required for motor-control and heater-control applications of the *Trigac* (Fig. 9-14).

CA3094 MOTOR-SPEED CONTROLLER

The three functional blocks of a motor-speed controller (Fig. 9-15) consist of the controller and motor circuit, the error detector, and

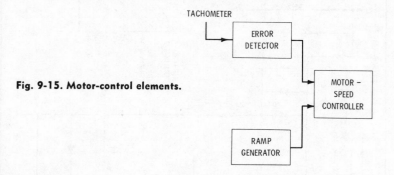

Fig. 9-15. Motor-control elements.

the ramp generator. The ramp generator provides a reference component in synchronism with the same power source that operates the motor. The error detector derives an error signal from the tachometer that is driven by the motor under control. The complete circuit is shown in Fig. 9-16.

The tachometer component, rectified and filtered, is applied to the input of a CA3080A IC, which is employed as a voltage comparator. The reference voltage at pin 3 is set by potentiometer R.

When the motor speed becomes high or low, there is an error voltage, E_1. If the motor speed is too low, the voltage at terminal 2 is less positive than at terminal 3, causing the terminal-6 voltage to go high. Conversely, a high motor speed results in a low voltage at terminal 6. This output voltage is applied to the phase comparator.

The ramp generator supplies a reference voltage to the comparator. It employs the CA3094A connected in a circuit that changes the pulsating dc voltage applied to its input to a sawtooth ramp voltage. This is accomplished by the rapid charging of capacitor C1 and the controlled slower discharge of the same capacitor.

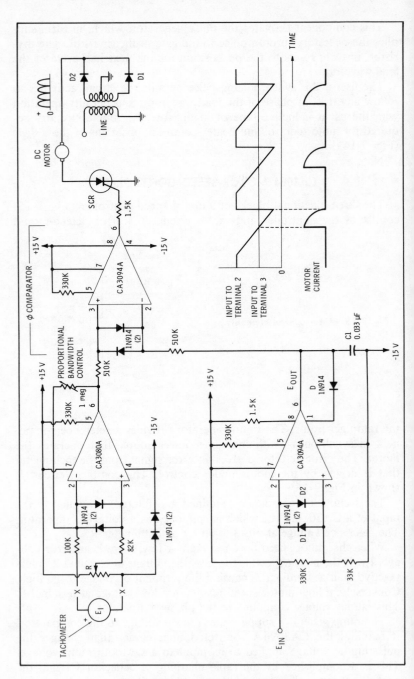

Fig. 9-16. Motor-speed control system.

The ramp voltage applied to terminal 2 of the comparator controls the motor current. The actual amount of motor current is set by the time duration of the positive signal applied to the semiconductor controlled rectifier (SCR) by way of pin 6. This actual conduction time is set by the dc voltage applied to pin 3 from the error detector. In fact, during an attempted change in motor speed, the error voltage changes accordingly and either steps up or steps down the level of the positive voltage at pin 6 in such manner that the motor current is made to maintain a constant motor speed.

Note from the waveforms that there is motor-speed current for whatever period of time the ramp voltage swings below average dc voltage level at terminal 3. If the dc voltage at terminal 3 rise in a positive direction toward the voltage at terminal 2, there will be a corresponding increase in the duration of the positive voltage at pin 6 and an increase in the duration of the motor current.

TEMPERATURE SWITCH

A simple electric/gas oven temperature control using an RCA CA3059 zero-crossing IC voltage switch is shown in Fig. 9-17. The sensor resistor has a negative temperature coefficient and, in conjunction with preset temperature-control resistor R_p, regulates the on and off intervals of a low-current SCR. This device controls a solenoid in an appropriate electric or gas oven. Any defect in the IC or SCR immediately removes power from the control.

OPTICAL ICs

In industrial systems, optical devices are used extensively for switching, limiting control, and evaluation. Such control devices and the op-

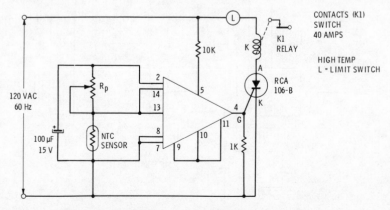

Fig. 9-17. An electric/gas oven-temperature controller.

tical detectors can be included in integrated circuits. The RCA CA3062 (Fig. 9-18) consists of a photosensitive detector, switching amplifier, and high-current output transistors. The power amplifier and its differential connection provide complementing outputs, which are a measure of the light that impinges on the integrated circuit through a glass port that opens on top of a conventional TO-5 IC case. The circuit arrangement is given in Fig. 9-19. In typcal application, a load for the photodetector is connected between pin 11 and common. An external connection is then made to connect the output of the detector to the input of the amplifier section at pin 1. Conventional differential amplifier and output circuitry follow.

Since the advent of integrated circuit PLLs, the phase-locked loop has become a practical means for motor-speed control (Fig. 9-20). In such an arrangement, the motor becomes the voltage-controlled oscillator (vco) of the loop. Attached to the motor shaft is an optical encoder which, in a typical case, provides an output of 36 pulses per revolution. This light-interruption technique provides a pulse train. These pulses are in turn applied to a divider and a phase comparator. A reference clock signal is applied to the same comparator, resulting in an error-voltage output. High frequencies are removed by an appropriate loop filter, and the control voltage is then applied to a motor-drive circuit, which sets the motor speed. Any attempted drift in motor speed results in an error voltage fed through the motor control circuit to hold the motor on proper frequency.

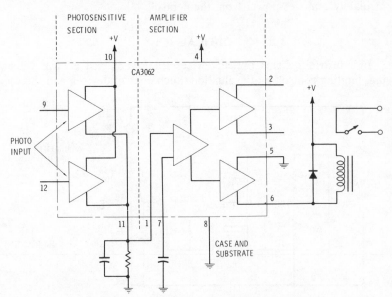

Fig. 9-18. Functional plan of RCA optoelectronic IC.

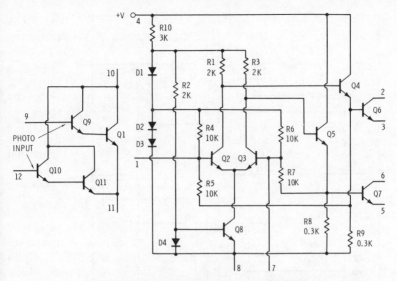

Fig. 9-19. Internal schematic of CA 3062 IC.

A light-operated servosystem using two optoelectronic ICs is given in Fig. 9-21. The servomotor rotates in either direction depending upon which optical detector is illuminated. Equal light level at both devices stops the motor. Note the external connection between pins 1 and 11. Manual override switches are also included.

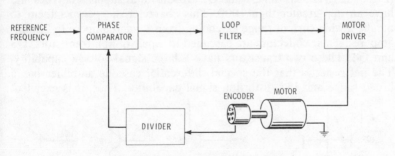

Fig. 9-20. PLL motor-control method.

HIGH-IMPEDANCE DC VOLTMETER

Linear and digital ICs are used extensively in modern test instruments. Their versatility, high stability, and small size are favorable attributes of test-equipment application. A high-gain transistor array and an operational amplifier are an ideal combination for such a meter.

The RCA CA3095 is a super-beta transistor array, including a differential cascode amplifier along with three independent transistors

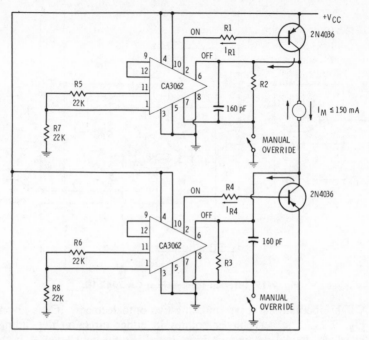

Fig. 9-21. Light-activated servomotor.

(Fig. 9-22). Transistors Q1 and Q2 are known as super-beta types and have an h_{FE} greater than 1000. This characteristic permits them to handle a wide current range extending from 1 microampere to 2 milliamperes. Their collectors are cascaded to a pair of npn transistors, Q3 and Q4. These two transistors have a large signal-voltage capability. The net result is that the special differential cascode amplifier has a broad-range and low-distortion signal capability. Thus, it is excellent

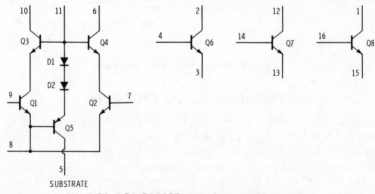

Fig. 9-22. RCA CA3095 super-beta transistor array.

for metering and also can be recommended for long interval timers, oscillators, and other sinusoidal and nonsinusoidal applications. A voltage-limiting network is connected between the bases of the output transistors and the input emitters.

A voltmeter circuit with an input resistance of 40 megohms is shown in Fig. 9-23. The array is connected in a bridge fashion with the voltage to be measured applied to base 9. Circuit balance and calibration are handled by the dc voltage applied to base 7. Two of the independent transistors of the array (Q7 and Q8) function as a constant-current source for the cascode differential amplifier.

The differential output of the amplifier is applied to the differential input of the CA3748 operational amplifier. Its output is connected to the 200-microampere indicating meter.

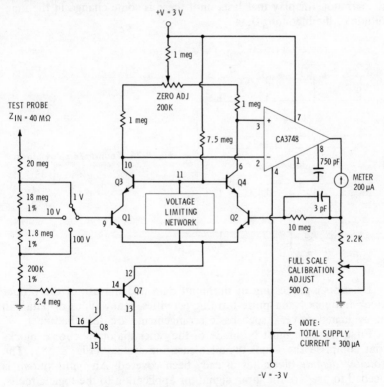

Fig. 9-23. High-impedance dc voltmeter.

DIGITAL DISPLAY INSTRUMENTS

The digital readout display has become commonplace in electronic test instruments. The digital display package consists of the readout light and decoder, the storage, and the counter chain (Fig. 9-24). Such a counter and display package can have a 3, 4, etc., readout capability, depending upon the range and accuracy desired.

A storage or memory segment is not an absolute necessity. However, such a unit does prevent the indicator lights from blinking as they roll from one reading to another.

Two types of information must be supplied to the digital display section. Under control of the clock, the data to be measured must be supplied to the input of the counter, signal A. After a sequence has been given to the counter, a transfer signal, B, instructs the counter to store and then display the status of the count. Finally, a reset signal, C, throws back the counting activity to zero, and a new cycle of operation is initiated. This activity repeats and repeats, maintaining a continuous display that lasts until there is some change in the magnitude of the incoming data.

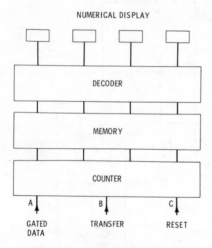

Fig. 9-24. Counter/display block.

By proper processing of the input data and the use of the proper clock frequency and pulse-forming activities, many types of data can be evaluated. Several such basic arrangements are described next.

First, mention must be made of the other major functional blocks that comprise the usual digital measuring system (Fig. 9-25). The *counter/display* block has already been covered. An *input system* is needed to prepare the input signal for application to the logic circuits. Some data require wave shaping and, for other types of measurements, a voltage-to-time conversion. A *master clock* is needed, usually crystal controlled, which, along with a frequency-divider chain, provides the proper timing for the various measurements to be made. Finally, a *control logic* and a *pulse generator* are needed. Here, the proper gating pulses are formed and the necessary pulses are passed to the counter. Other logic functions must also occur, depending upon the measurements to be made.

Frequency-Period Measurements

A method of frequency measurement is shown in Fig. 9-26. The input sine-wave frequency is first converted to a corresponding squared

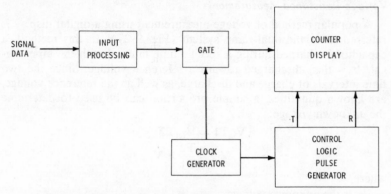

Fig. 9-25. Functional plan of a digital measuring instrument.

wave with the same repetition rate. It is supplied to a gate that, in turn, feeds signal to the counter/display block (Fig. 9-26A). However, this information is gated on and off by a precise gating pulse from the clock generator. This pulse of precise duration turns on the count gate for a specific time interval. It might be 10 milliseconds or 100 milliseconds, depending upon the frequency range to be measured and the design particulars of the counter. The same signal is applied to the pulse generator of the control logic block.

In operation the counter will now count the number of cycles that occur during the clock period. At the end of the period, the edge of the clock signal causes the pulse generator to initiate two successive pulses. The first one is the transfer pulse, which causes the counter unit to store and then display the state of the individual count in each channel. A second, or reset, pulse then leaves the pulse generator, resetting the count section to zero prior to the start of the next cycle of operation. These cycles repeat and maintain a steady reading. Readings change whenever the frequency of the input signal varies.

A similar arrangement is used to measure the period of an applied input wave (Fig. 9-26B). As shown in the accompanying waveforms, the clock frequency is now made higher than the frequency of the input wave. The waveshaping circuit and a binary counter produce a pulse output with a duration corresponding to one period of the input wave to be measured. During this interval of time, the clock pulses are applied to the counter. Since they are of a known frequency, they can be used to take a time measurement (period) of the input wave. This information is displayed as a frequency. However, it can be converted to a period using the simple equation:

$$\text{Period} = \frac{1}{\text{Frequency}}$$

Voltage-Resistance Measurements

A popular method of voltage measurement using a digital display is referred to as the dual-ramp system (Fig. 9-27). In this method, a capacitor is charged during a specific time interval with the input signal. It is then discharged toward a reference voltage. Since the two time intervals of charge and discharge, as well as the reference voltage, are known quantities, a simple proportion can be used to determine the unknown voltage:

$$V_{in}T1 = V_{ref}T2$$
$$V_{in} = \frac{T2}{T1} \times V_{ref}$$

where,

V_{in} is the unknown voltage,
V_{ref} is a known reference voltage,
T1 is charge time,
T2 is discharge time.

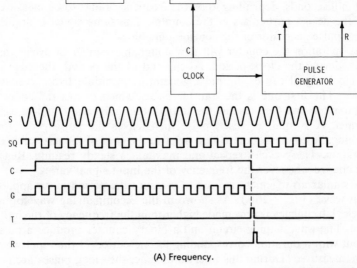

(A) Frequency.

Fig. 9-26. Frequency

Inasmuch as the digital display can be used to display time interval, a suitable calibration can permit the direct readout in voltage.

At the beginning of the measurement, the capacitor charges for an interval of time set precisely by a timing pulse applied to the control logic circuit. This pulse is derived from the clock generator by passing the clock signal through the counter.

After the calibrated interval, a reference voltage is applied to the input of the integrator. The capacitor now discharges. However, its rate of discharge is greater than its rate of charge because the reference voltage is made higher than the input voltage to be measured. Therefore, in the time, T2, the capacitor has discharged to 0. This is verified by a comparator, which then generates a pulse leading edge that is applied to the control logic circuit, and the transfer pulse initiates the store and display reading. The actual reading displayed depends upon the time intervals during which pulses were being applied to the counter. Thus, the T2/T1 ratio is recorded. This display gives a direct measure of the input voltage in terms of reference voltage. When all activities return to 0, the cycle repeats, providing a continuous display.

Although measuring instruments use more digital ICs than linear types, there are a variety of operational amplifiers and waveform generators, as well as shaping circuits, to be found.

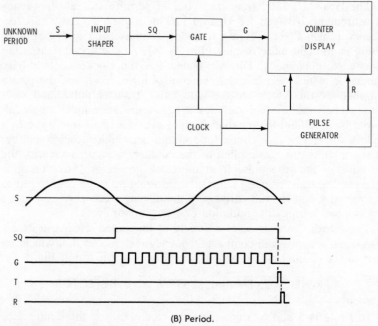

(B) Period.

and period measurements.

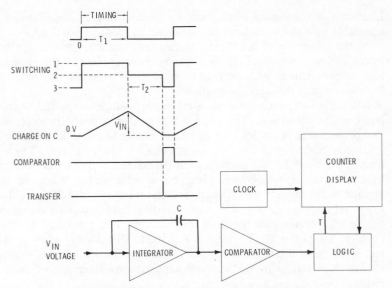

Fig. 9-27. Voltage and resistance measurements.

HEATH FREQUENCY COUNTER

The Heath IB-1102 frequency counter permits accurate frequency measurements between 1 hertz and 120 megahertz. To cover this frequency range, there are gate times of 1 millisecond and 1 second. There is an eight-place display (Fig. 9-28). At the upper right is the counter/display block. Directly beneath is the clock and time-base generator, which also includes the control logic. Note that the pulses supplied by this block are the gate pulse, transfer pulse, and reset pulse. The input system includes a high-impedance input stage followed by a Schmitt trigger that changes over the input sine wave to a square wave. The gate circuit follows and permits a specified number of input pulses to be supplied to the counter in accordance with the duration of the gate pulse from the clock generator. After the gate interval, the transfer pulse allows the information to be transferred to the memory and the tube drivers of the display section. A reset pulse follows and permits the initiation of a new count cycle.

The Heath Company uses a variety of integrated circuits in its design of an accurate and compact frequency meter. The ICs, which are part of the counter circuit board, can be seen at the top of Fig. 9-29.

PROJECT 5: FM I-M SYSTEM AND DETECTOR

General

In Projects 5 and 6, a complete fm receiver using three integrated circuits is constructed. The first project involves the 10.7-MHz i-f

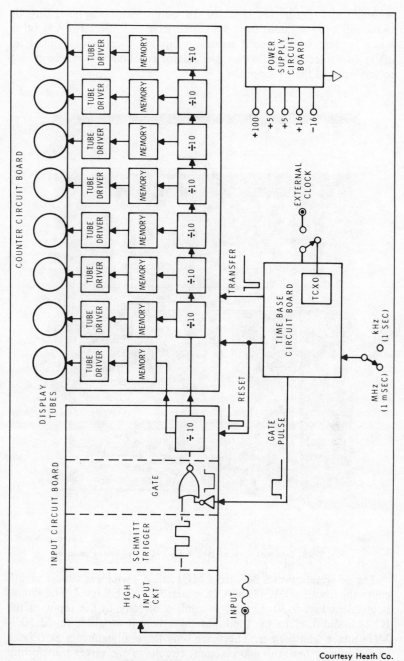

Courtesy Heath Co.

Fig. 9-28. Block diagram of a frequency counter.

333

amplifier and quadrature detector. Its output can be applied directly to the input of the audio amplifier constructed in Project 1. (Any other audio module will suffice if it has an input sensitivity of at least 30 mV.) In Project 6, an fm mixer-oscillator combination will be added to the input of the i-f amplifier.

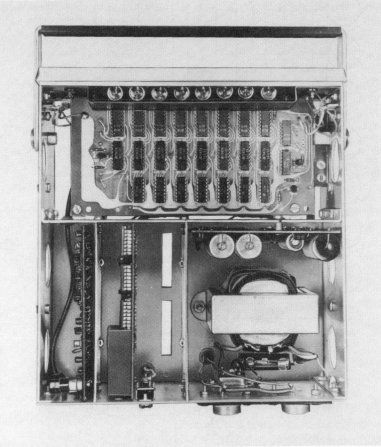

Courtesy Heath Co.

Fig. 9-29. Internal view of Heath IB-1102 frequency counter.

The i-f system uses a Motorola MC1357 IC, which is almost identical to the Sprague ULN2111A IC described in Chapter 7. The circuit is given in Fig. 9-30. Output is applied directly to the input of the IC-21 audio module. On the input side, there is a double-tuned 10.7-MHz mixer transformer. There are only three adjustments associated with the i-f system, the two resonant circuits of the mixer transformer and the quadrature coil, L2.

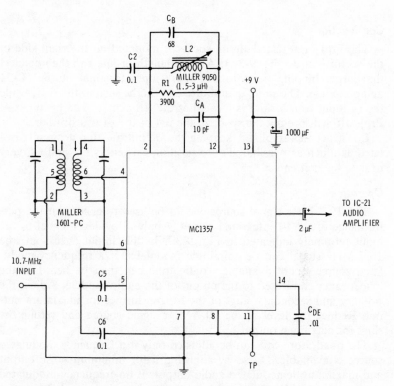

Fig. 9-30. An fm i-f system.

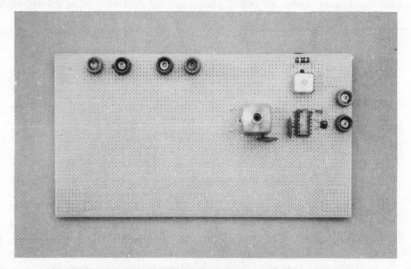

Fig. 9-31. Vector board showing mounted parts for fm i-f system.

Construction

The fm i-f integrated-circuit socket is mounted on the right side of the vector board (Fig. 9-31). A short length of line can then connect the output binding posts to the audio input terminal of the IC-21 audio module. The small quadrature coil can be seen to the rear, while the i-f input transformer is located at the left center. The fm tuner of Project 6 will occupy the space to the left of the i-f transformer.

Construct the amplifier, keeping the output and input circuits isolated as much as possible. Try to position capacitors C2 and C5 very near to the coil ends.

Operation

The 10.7-MHz signal source should be used to check out the performance of the i-f integrated circuit. Apply a low-level signal to the input terminals indicated in Fig. 9-30. In this initial check-out, the 10.7-MHz signal can be amplitude modulated if a frequency-modulated source is not available. Audio tone can then be heard in the loudspeaker connected to the output of the audio module. Adjust the primary and secondary slugs of the i-f transformer for maximum output. Reduce the level of the 10.7-MHz signal source and retune the slugs for maximum output.

The quadrature coil can be adjusted only if a frequency-modulated source is available. Do so by adjusting it for minimum noise output and maximum demodulated audio output. If no frequency-modulated source is available, do not change the setting of the coil. It can be adjusted later when an actual frequency-modulated broadcast signal is being received.

PROJECT 6: THREE-IC FM RECEIVER

General

A complete and reasonably sensitive fm receiver can be constructed by adding just one more integrated circuit to the i-f system and audio module of Project 5. An RCA CA3005 IC can be made to perform as a capable fm tuner consisting of input amplifier, mixer, and local oscillator. This is a 12-pin device mounted in a TO-5 case. The device was described in Chapter 7 (refer to Fig. 7-22 and associated text).

If greater sensitivity and selectivity are desired, a two-stage fm tuner can be built (Fig. 7-23). However, the single-stage device provides adequate performance although it has some image response. Separate mixer and local-oscillator tuning capacitors are incorporated (Fig. 9-32), instead of using a gang tuning capacitor. This permits one to tune the input more precisely and aids in tuning out any annoying image response that may be encountered in the reception of a particular fm station. If a 455-kHz i-f is preferred, data can be found in Fig. 7-27.

Construction

Construct the fm tuner on the left side of the vector board. Keep the mixer and local-oscillator circuits as isolated as possible and use short-lead connections. Also, keep the 10.7-MHz trap away from the other two coils. In the test construction, the actual integrated circuit was wired permanently into the circuit, making use of vector terminal connectors. Of course, when permanent wiring is done, a small heat sink is used when soldering directly to IC leads.

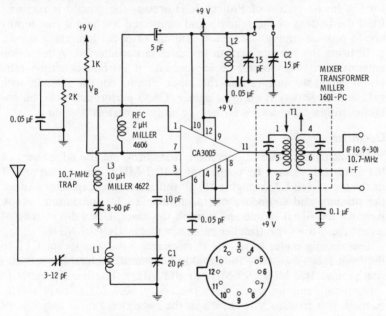

Fig. 9-32. Mixer-oscillator of the fm tuner.

The rf tuned coil consists of three turns of No. 22 enameled wire (center-tapped) mounted on a ¼-inch diameter vhf coil form (Miller 4500-4). Spacing the turns can be used to set the frequency range over which the 20-pF variable capacitor will tune. The slug in the coil is also an aid in tuning. Using a dip meter, it was no problem to obtain a tuning range between 75 MHz and 115 MHz.

The local-oscillator tuning is accomplished with two 15-pF variable capacitors, one of which can be taken in or out of the circuit by using a jumper. These variable capacitors are mounted on a single insulated shaft, finalizing in a vernier tuning knob. When only one capacitor is in the circuit, the high-frequency end of the fm broadcast band can be fine-tuned. The two capacitors in parallel are used to tune the low end of the fm broadcast band.

Coil L2 itself consists of 2½ turns of No. 22 enameled wire wound on a vhf toroid form (black T-50-10). Note that the bottom end of the coil is connected to +9 volts. Be certain that the 0.05-μF capacitor is connected directly between the bottom of the coil and the nearest ground point (preferably the rotor side of the tuning capacitor).

The oscillator frequency can be checked out with a dip meter. However, the oscillator must be placed in operation to obtain an acceptable indication on most types of dip meters when using a toroid coil.

The mixer transformer is the same one mounted on the pegboard for the fm i-f system of Project 5. However, the ground is removed from the bottom of the primary and connected to the +9-volt supply line to provide supply voltage for terminal 11 of the integrated circuit.

In using the fm receiver, an fm antenna is preferred. A television antenna is a poor second choice because it introduces strong television broadcast images into the input circuit. An fm antenna with gain and directivity is an ideal choice. Good performance can be obtained within 50 miles, or farther, of an fm broadcast station.

Operation

In placing the tuner in operation, first check out the adjustment of the mixer transformer by connecting a 10.7-MHz signal to terminal 1 or the top of coil L3 through a 10-pF series isolating capacitor. Adjust the primary and secondary of transformer T1 for maximum output. Remove the signal source and turn off the tuner. Use a dip meter and adjust the 8-to-60-pF trimmer capacitor to exactly 10.7 MHz.

Use the dip meter to set the rf resonant circuit (capacitor C1) to the frequency of a strong fm signal in your area, preferably some station around 100 MHz (99 MHz to 102 MHz). Use the dip meter to set the oscillator frequency to a frequency 10.7 MHz lower in frequency. If a frequency is chosen in the suggested range, only one of the oscillator capacitors need be connected in the circuit.

Now connect the fm antenna to the tuner. Turn on the receiver. The station, or at least background noise, should be heard in the output. Adjust the oscillator variable until the signal is tuned in. Tune in the signal very carefully. You can now adjust the quadrature coil of the fm i-f system for minimum background noise and maximum audio output.

Tune for other stations at the high end of the fm broadcast band. Incoming signals can be peaked by adjusting rf tuning capacitor C1. Next, connect the two oscillator capacitors in parallel and set capacitor C1 for tuning at the low-frequency end of the fm broadcast band. Now you can tune over the low end of the band, using the oscillator variable and adjusting the final peaking with capacitor C1.

Table 9-1. Parts List for Projects 5 and 6

Qty	Description
1	IC-21 audio module
1	IC socket, 14-pin in-line
1	MC1357 Motorola integrated circuit
1	CA3005 RCA integrated circuit
1	Vector board, 4½" × 8½"
1	20-pF variable capacitor
2	15-pF variable capacitors (ganged)
1	Vernier tuning dial
1	3- to 12-pF trimmer capacitor
1	8- to 60-pF trimmer capacitor
1	10-μH rf coil (Miller 4622)
1	3-foot length No. 22 enameled copper wire
1	Toroid coil form T50-10
1	Ceramic coil form, ¼" in diameter (Miller 4500-4)
1	10.7-MHz mixer i-f transformer (Miller 1601-PC)
1	1.5 to 3-μH quadrature coil (Miller 9050)
1	2-μH RFC (Miller 4606)
6	Binding posts
1	1000-μF 16-volt electrolytic capacitor
1	2-μF nonpolarized capacitor
4	0.1-μF disc capacitors
3	0.05-μF disc capacitors
1	0.01-μF disc capacitor
1	68-pF disc capacitor
2	10-pF disc capacitors
1	5-pF disc capacitor
1	1K, ½-watt resistor
1	2K, ½-watt resistor
1	3.9K, ½-watt resistor

10

Two-Way Radio

Except for the high-powered stages of the transmitter, integrated circuits are feasible in all sections of modern radio transmitters, receivers, and transceivers in the marine, aviation, land-mobile and amateur radio services. Presently, the discrete field-effect and bipolar transistors outnumber the integrated circuits in modern gear. However, ratios will favor the IC in the not-too-distant future. Even in the realm of radio-frequency power amplification, a limited number of IC devices have already been developed and marketed. Two-way radio with its emphasis on compactness and lightweight is a natural for the integrated circuit.

HYBRID RF POWER AMPLIFIER

Motorola has developed two hybrid IC power amplifiers for the land-mobile radio service. These operate in the uhf spectrum between 400 and 470 MHz, with a dc supply voltage of 12.5 volts. Output power is a minimum of 13 watts, with an input driving power of only 150 milliwatts (Fig. 10-1). A second device delivers an output of 7.5 watts, with a driving power of only 100 mW.

The input impedance is 50 ohms for both devices, and they are capable of operating with as high as a 20-to-1 load mismatch without damage.

There are 7 pins, 3 of them grounds. Two pins are provided for the rf input and output. There is a terminal for application of the +dc voltage and a terminal that can be used to regulate power gain using an external network.

The device is less than 3 inches long and less than 1 inch wide. These dimensions, along with the capability of the device, emphasize

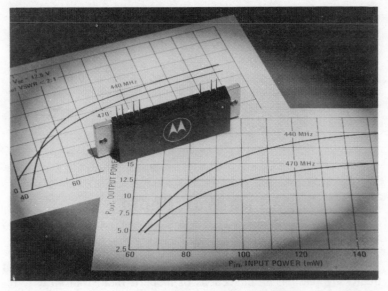

Fig. 10-1. Hybrid IC rf power amplifier.

the inevitable widespread application of integrated circuits in two-way radio gear.

SYSTEMS AND SUBSYSTEMS

It is not unreasonable to expect that complete low-power transmitters, receivers, and even transceivers will eventually be packaged in a single hybrid and/or monolithic enclosure. The Lithic complete a-m transmitter (Fig. 10-2) was described previously in Chapter 6. Component values are shown for a complete 27-MHz a-m transmitter. This unit provides a 100-milliwatt output, and the values shown will operate on the Citizens band and the nearby 10-meter amateur band. Lithic now makes available a 250-milliwatt output version of this integrated circuit.

The Lithic LS371 IC (Fig. 10-3) is a differential/cascode amplifier for high-frequency application. It also includes the necessary biasing and compensating diodes.

For a Citizens band or 10-meter ham application, the LS371 can be used as the front end of a well-performing receiver (Fig. 10-4). In the cascade connection, the input signal is applied to the base of the constant-current transistor. One side of the differential pair operates as a crystal-controlled oscillator. Oscillations mix with the incoming signal to produce a difference frequency in the collector circuit of the second

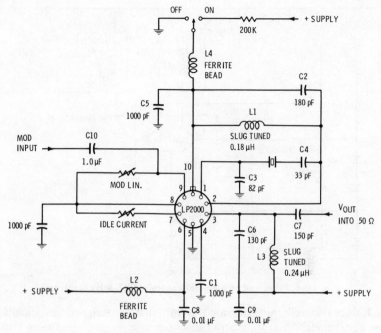

Fig. 10-2. A milliwatt a-m transmitter.

transistor of the differential pair. Output is removed by way of the 455-kHz i-f tranformer.

The same transistor can be used as a stable high-gain emitter-coupled vhf amplifier (Fig. 10-5). By proper selection of coil dimensions for L1 and L2, this amplifier can be made to operate in the fm broadcast band, as well as in the 2- and 6-meter amateur bands. Input and output matching are handled by the split-capacitor arrangements.

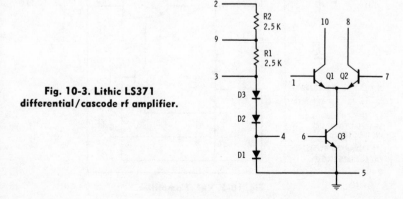

**Fig. 10-3. Lithic LS371
differential/cascode rf amplifier.**

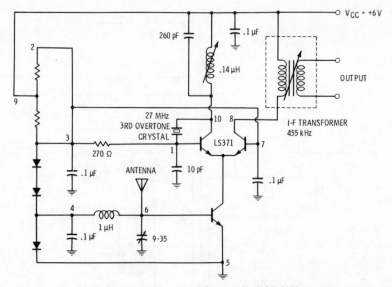

Fig. 10-4. Front end of a crystal-controlled 27-MHz receiver.

Lithic also sells an integrated circuit that is designed specifically for agc/squelch/vox application in communications transmitters and receivers. A typical vox circuit and a speech-compression amplifier

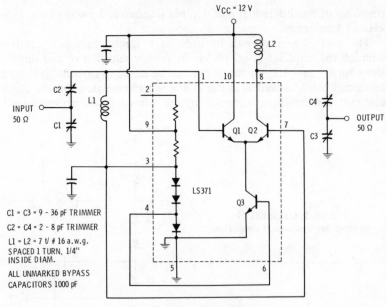

C1 = C3 = 9 - 36 pF TRIMMER
C2 = C4 = 2 - 8 pF TRIMMER
L1 = L2 = 7 t/ # 16 a.w.g.
SPACED 1 TURN, 1/4"
INSIDE DIAM.
ALL UNMARKED BYPASS
CAPACITORS 1000 pF

Fig. 10-5. Vhf rf amplifier.

are shown in Fig. 10-6. The purpose of a vox circuit is to automatically turn on the transmitter when the operator speaks into the microphone. As soon as he stops talking, the transmitter-receiver switches back to receive automatically.

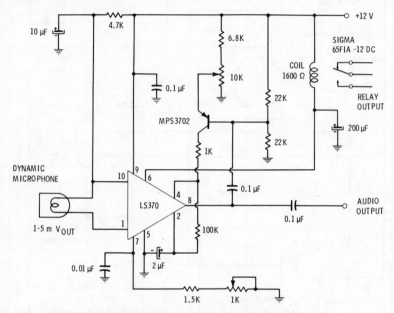

Fig. 10-6. Simultaneous voice-operated switch (vox) and speech-compression preamplifier.

Internal circuitry (Fig. 10-7) shows the differential-amplifier input applied between pins 1 and 10 at top left. Output from one pair of collectors is applied to the separate bases of the dual-differential pair shown at the top center. Note that their common bases connect back to the control circuitry at the bottom center.

Control-current components applied at positions A, B, or C are able to change the gain of the dual-differential pairs and thus control the level of the signal, which is emitter-coupled to the output transistors.

A second pair of collector outputs from the input differential amplifier is applied to the dc amplifier system at the lower left in Fig. 10-7. Output at pin 6 is a dc component that can be used for a receiver squelch circuit or a transmitter vox circuit.

Refer now to the vox circuit of Fig. 10-6. Note that the microphone is connected between pins 1 and 10. Observe that a dc current component at pin 6 passes through a relay coil. When the relay is energized, the associated power switch turns on the transmitter. Thus, by

345

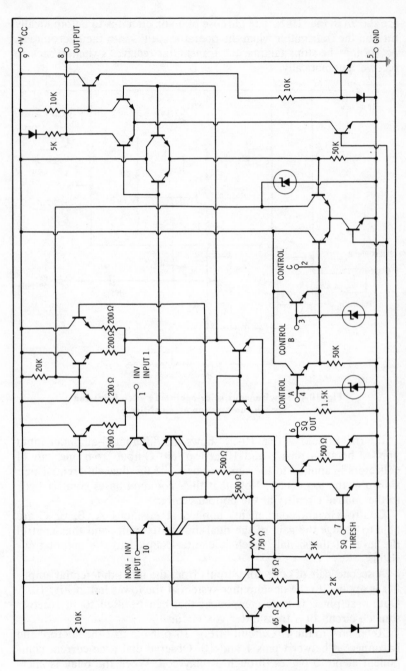

Fig. 10-7. Lithic LS370 vox/squelch/agc integrated circuit.

346

speaking into the microphone, a dc component at pin 6 turns on the transmitter. When the operator stops speaking, there is a momentary delay and then the unit switches back automatically to the receive position.

The audio output is derived at pin 8. A component of the audio is coupled back to pin 4 of the integrated circuit by way of the $0.1-\mu F$ capacitor and a discrete bipolar transistor. Note that this is one of the control inputs of the integrated circuit (Fig. 10-7). This component is rectified and develops a dc voltage that is used to regulate the gain of the audio amplifier. Therefore, if the audio output is of a level too high for proper operation of the transmitter, the gain of the integrated circuit is reduced automatically to maintain a reasonably constant audio output at pin 8. This manner of operation is referred to as speech compression. It compensates for different voice levels and maintains the proper modulation level for the transmitter without the use of a speech-gain-control adjustment.

Complete FM I-F Subsystem for Communications

The RCA CA3089E IC (Fig. 10-8) can serve as a complete receiver subsystem for a communications package. It includes a three-stage fm i-f amplifier/limiter channel, plus individual signal-level detectors for each stage. There is a doubly balanced quadrature fm detector that can be used with either a single-tuned or double-tuned detector coil. The quadrature detector is followed by an audio amplifier with a typical 400-mV output level. The quadrature detector also supplies drive to an automatic frequency-control amplifier, whose output can be used to hold the local oscillator on correct frequency.

The level-detector stages supply signal for a tuning-meter circuit and a delay agc component for the rf amplifier of the tuner. A level detector is also associated with the quadrature detector. It supplies a drive component to the squelch system. The squelch output, in turn, is supplied to the muting stages of the audio amplifier. Such a squelch circuit silences the output of the receiver when there is no incoming signal above a level set by the muting sensitivity control. A typical single-tuned detector circuit is shown in Fig. 10-9. Practical values are given for operation in the 10.7-MHz spectrum.

The LM373 is a communications IC made available by National Semiconductor (Fig. 10-10). This device includes an i-f amplifier/limiter and the necessary internal circuitry to provide various demodulation modes. Suitable external switching permits the device to operate as an a-m, fm, or ssb/cw demodulator. Bandpass shaping is accomplished with suitable external discrete LC or network components. The bandwidth extends up to 30 MHz.

The row of stages at the top of the block diagram includes an amplifier/limiter, agc, gain-control input and a second gain circuit.

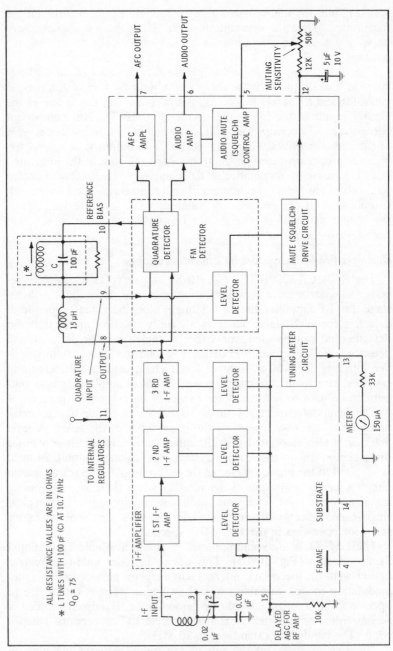

Fig. 10-8. RCA CA3089E integrated circuit for an fm i-f system used in communications equipment.

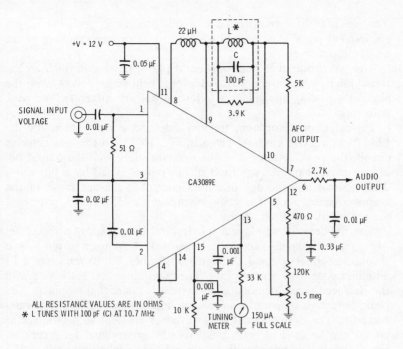

Fig. 10-9. Complete external circuit for single-tuned detector.

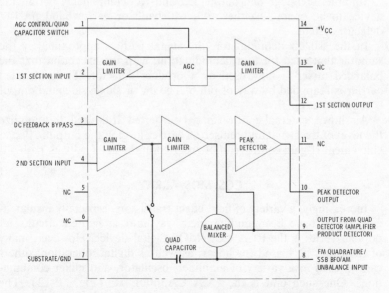

Fig. 10-10. Internal plan of National Semiconductor LM373 IC.

The signal to be demonstrated is applied to pin 2, with the output removed at pin 9. Agc control is obtained by proper sampling of the audio-output level.

The pin 9 output is reintroduced at pin 4, passing through an external filter circuit on the way. Additional gain stages follow, with the signal then being applied to a following balance mixer and/or peak detector.

Typical circuits for a-m, fm and sideband demodulation for the LM373 in a TO-5 case are given in Fig. 10-11. In the a-m detector circuit, the peak detector does the demodulation, with the output being removed at pin 8. A portion of this output is fed back through a 3.9K resistor and the agc threshold control to the agc input of the amplifier chain, pin 1. Typically, when using a 455-kHz i-f frequency, the sensitivity is 10 microvolts.

The fm detector employs a quadrature demodulator, with the fm i-f signal being applied to pin 2, being picked up again at pin 9, and then transformer-coupled through transformer T1 to the second i-f amplifier system by way of pin 4. The quadrature coil associated with the balanced-mixer demodulation system is connected to pin 6. The peak detector is inoperative in this method of demodulation for wideband fm. The output signal is removed at pin 7. The peak-detector circuit can be used to limit bandwidth in narrowband fm reception. Good output is obtained with only a ±5-kHz deviation at either 455 kHz or 10.7 MHz.

Suitable interstage and output LC and RC components permit the device to be operated either as a wideband or narrowband fm i-f/detector system.

In the ssb/cw demodulator, the signal path is approximately the same as that used for a-m demodulation, with the exception that the balanced mixer is now used as a product detector. The reinserted carrier is reapplied by way of pin 6. The cw or single-sideband output is removed at pin 7.

As shown, external controls may be inserted, if desired, to minimize the level of the signal and reinserted carrier level in the output. Also, a convenient rf-gain control can be inserted in the agc feedback path.

COS/MOS ARRAY

Increasingly, a variety of field-effect transistors, especially insulated-gate types, have been finding their way into integrated circuits. A popular series is the RCA COS/MOS digital device. However, many of these can be biased for linear, as well as digital, operation; others can be used in a variety of amplifier, oscillator, and mixer combinations. One such unit is the RCA CA3600E IC (Fig. 10-12). The practical device with pin-out connections for applying external com-

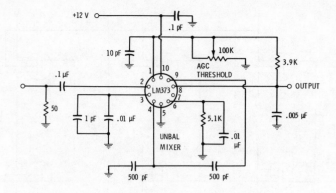

(A) A-m demodulator.

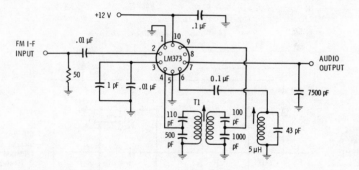

(B) Fm demodulator.

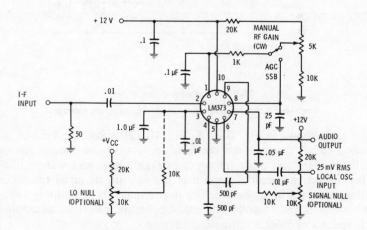

(C) A ssb/cw demodulator.

Fig. 10-11. Practical demodulator circuits.

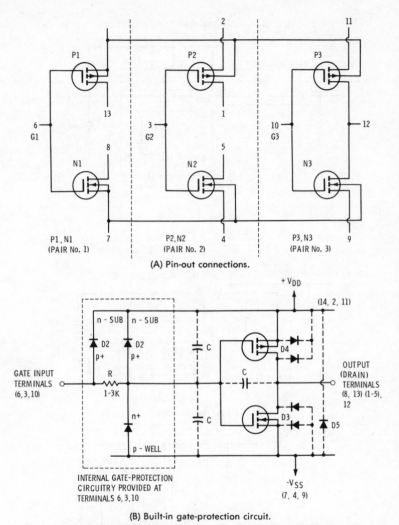

(A) Pin-out connections.

(B) Built-in gate-protection circuit.

Fig. 10-12. RCA COS/MOS CA3600E integrated circuit.

ponents is shown in Fig. 10-12A; Fig. 10-12B shows the built-in protection circuit that protects the insulated gates from damage by high static voltages and other improper external signal and voltage levels. This protective circuit is present at the gate inputs of all three transistors mounted in the device. The device operates at a supply voltage of between 3 and 15 volts, and useful operation can be obtained up to a frequency of 5 MHz in untuned circuits.

As an oscillator, the circuit can be made to operate at considerably higher frequencies (Fig. 10-13). The device can be operated in crystal-controlled or in vfo fashion, with the insertion of an appropriate LC tunable resonant circuit. In the circuit a sine-wave output can be obtained. Note that only one of the three FET pairs in the array is used in the circuit.

Each FET pair (Fig. 10-14) includes two complementary linear amplifier circuits. Linear biasing of the complementary pair is handled by resistor R_b connected between pins 10 and 12, giving due con-

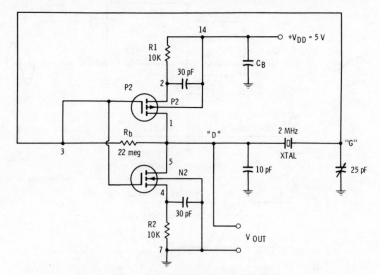

Fig. 10-13. A COS/MOS crystal-oscillator circuit.

sideration to the value of the source resistance, R_s. Typical voltage gain and bandwidth as a function of supply voltage are given in Fig. 10-14B. The curve assumes a value of 22 megohms for R_b and a 50-ohm signal source.

SIDEBAND MODULATORS

A popular type of integrated circuit is the doubly balanced differential-amplifier combination. This is used extensively in modulator and demodulator circuits, especially in double-sideband and single-sideband generators.

The Motorola MC1596G IC previously described is one such type (Fig. 10-15). This complete generator can be used for forming a double-sideband signal (both sidebands present with the carrier suppressed) over a wide frequency range, depending upon crystal frequency or vfo input frequency. The same arrangement with the use of a 9-MHz crystal and a 9-MHz sideband filter can be used to generate either an upper or lower 9-MHz single-sideband signal.

The carrier generator is shown at the top left (Fig. 10-15) and uses an FET crystal oscillator followed by an isolation buffer stage. If desired, a vfo signal can be applied to the input gate. For this manner of operation, one need only remove the crystal from its socket.

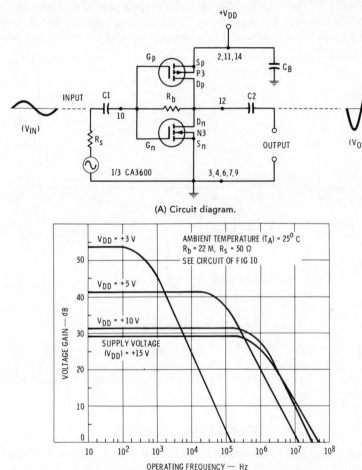

(A) Circuit diagram.

(B) Voltage gain and bandwidth.

Fig. 10-14. A COS/MOS linear amplifier.

The modulating wave is applied to the lower left integrated circuit using the HEP 580 as a two-stage audio amplifier. Its output is transformer-coupled to the input of the double balanced modulator. The carrier component is applied to pin 8.

A double-sideband signal is removed at pin 6 and applied to the two-stage bipolar transistor amplifier. These two stages are used as a

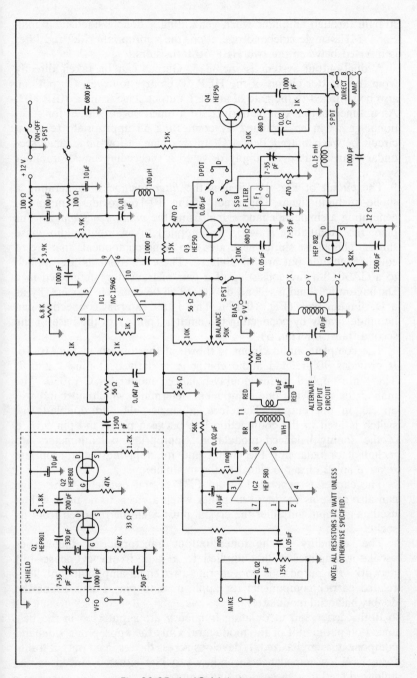

Fig. 10-15. An IC dsb/ssb generator.

355

straight-through amplifier when generating a double-sideband signal. For 9-MHz single-sideband operation, the appropriate sideband filter is inserted between the two HEP 50 transistors.

A dual-output system is included. Output can be taken directly from the emitter circuit of the HEP 50 emitter follower. Output can also be switched to the gate of an FET output stage using a HEP 802. An untuned output can be derived, or a tuned output circuit for additional gain can be arranged by connecting an appropriate resonant circuit between output B and C. In this plan, a low-impedance secondary can be used to supply drive to a succeeding linear-amplifier system.

The circuit of Fig. 10-15 uses the *filter method* of generating a single-sideband signal. The circuit of Fig. 10-16 provides a means of generating a single-sideband signal using the *phasing method*. In this plan, the modulating wave is first applied to an input audio-phase shifter. This phase shifter generates two audio components that are of equal magnitude but are 90° related. These components are applied to a CA3018 array consisting of a pair of transistors connected in a Darlington pair and a second duo that can be connected externally as a Darlington pair (Fig. 10-17). This latter external connection can be made simply by connecting the emitter (pin 4) to the base of the second transistor (pin 6).

The complete audio circuit is shown at the top left of Fig. 10-16. It provides 90°-related audio components derived at pins 1 and 7. These are applied to the doubly-balanced modulator CA3050. The manner of supplying one component determines which sideband will appear in the output. Therefore, a single-pole and double-throw switch is used in the conjunction line between pin 7 and pin 5 or 8 of the doubly balanced modulator. Appropriate potentiometers are included for balancing the circuit and for applying the appropriate relative levels of audio signal to the modulator.

The carrier is generated by an FET crystal oscillator. The stage can also serve as the input amplifier when a vfo is to be used. When used as an input amplifier the crystal must again be removed from its socket.

The secondary of the tuned output transformer of the carrier oscillator/amplifier is first applied to a radio-frequency phase-shift network. Its purpose is to generate two equal-amplitude but 90°-related carrier components for application to pins 1 and 13 of the doubly balanced modulator.

Both carrier and modulating frequency are suppressed in the balanced-output circuit of the modulator. Only the appropriate sideband component is emphasized. It develops across the resonant output transformer. A low-impedance secondary provides either balanced or unbalanced feed to a succeeding linear amplifier.

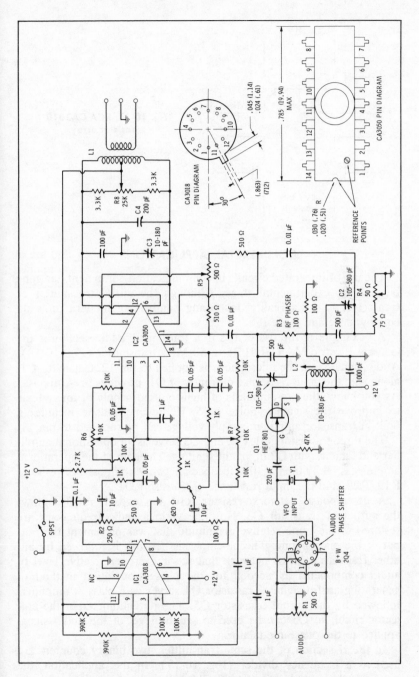

Fig. 10-16. A 160-meter phasing-type sideband generator.

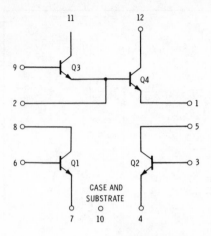

Fig. 10-17. RCA CA3018 transistor array.

COMMERCIAL APPLICATIONS

The use of integrated circuits in communication equipment has gone beyond the experimental stage. Although the all-IC transmitter or transceiver is not a reality, increasing numbers of modern units do employ more and more ICs.

An example of the use of ICs in a transmitter audio section is the General Electric MASTR II 25-to 50-MHz transmitter (Fig. 10-18). Audio signal from the microphone is applied to pin 12. Capacitor C1, along with the input resistance of transistor Q1, provides pre-emphasis. As you know, the pre-emphasis of highs is used widely in fm systems to improve the signal-to-noise ratio. If the microphone includes a built-in transistor amplifier, supply voltage can be supplied to pin 11.

The three transistors Q1, Q2, and Q3 serve as an operational amplifier and limiter, with Q3 doing the limiting. The gain of the operational amplifier is set by the negative feedback network composed of resistors R19, R20, and the input resistance of transistor Q1.

A de-emphasis network, resistor R10 and capacitor C3, follows the output of transistor Q3. Now the various audio frequencies are restored to the same relative magnitude that was present at the output of the microphone. The de-emphasis network is followed by another resistor-capacitor network that sets the proper relative level of audio components as needed for the phase-modulation method of generating an fm signal. Transistor Q4 serves as a class-A amplifier, followed by the output transistor Q5. The potentiometer in the collector circuit of Q5 can be used to set the level of the audio signal applied to the phase modulator.

In the rf section of the same transmitter, two binary counters are used as a frequency divider (Fig. 10-19). In this application, the output of the crystal-controlled oscillator that determines the eventual

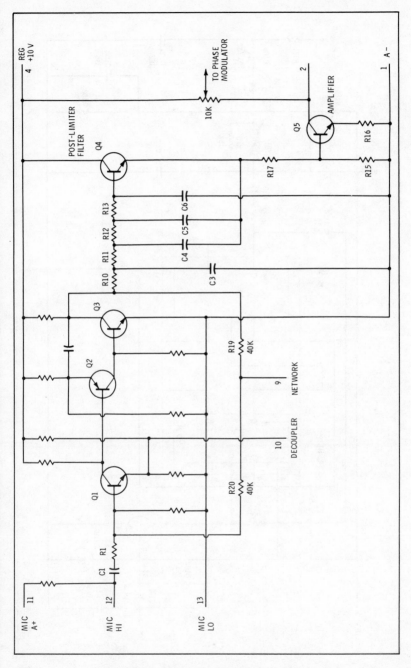

Fig. 10-18. Speech amplifier and processor.

359

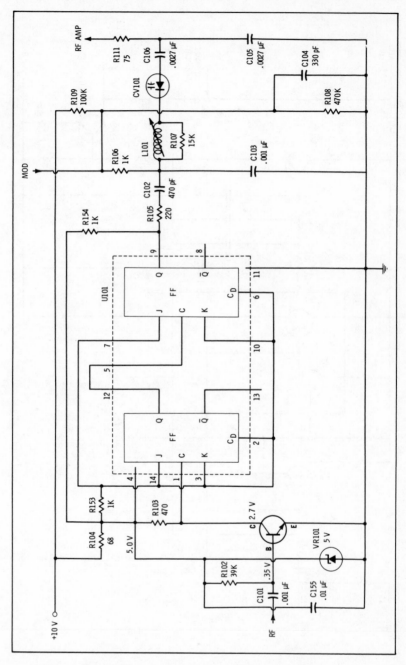

Fig. 10-19. Frequency divider and phase modulator.

360

frequency of the operation of the transmitter is applied to the base of the bipolar transistor at the input of the divider. This input transistor and associated diode assist in shaping the oscillator wave for application to the 4-to-1 divider. In fact, the input transistor is cut on and off by the carrier component.

As it is turned off on each cycle, its drop in collector voltage changes the state of the first digital flip-flop. Two oscillator cycles cause the flip-flop to switch one complete cycle from state 0 to state 1 and back to state 0. Therefore, there is one cycle of output for each two cycles of input. The signal is then applied to a second flip-flop binary counter, producing a total division of 4 (2×2).

The square-wave output of the divider is converted to a sine wave by the input resonant circuit of the voltage-variable capacitor diode, CV101. Note that the audio-modulating wave is applied to the same input circuit. As a result, a frequency-modulated wave appears at the output being generated by the indirect phase-modulation process. Resistors R108 and R109 provide the dc biasing for the CV101 varactor.

This manner of generating an fm signal is attractive in that the crystal-controlled carrier frequency is first reduced in frequency by a factor of 4. The modulation occurs at this frequency. However, the transmit frequency is much higher and a succession of multiplier activities must follow. For any multiplication of the fm wave, there is also a multiplication of the frequency deviation. Thus, it is possible to obtain a substantial linear-frequency deviation if the modulation occurs at a relatively low carrier frequency.

AERONAUTICAL RADIOCOMMUNICATION

Four important electronic aids to aircraft travel are radiocommunication, navigation, traffic control, and landing. Most modern-day aeronautical radio activity occurs in the frequency spectrum between 108 and 136 MHz. Radio navigation uses the spectrum between 108 and 118 MHz; air-traffic control and communication, 118 to 136 MHz. Scattered throughout this spectrum are frequencies assigned to both aircraft and aeronautical ground stations. For example, in flying a private aircraft, you will find the frequencies of the various ground stations given on navigation maps and/or charts. These aeronautical ground stations monitor certain aircraft frequencies, and you can quickly establish contact en route by setting your aviation radio to an appropriate frequency.

Many aviation radio units have dual-reception facilities. Thus, it is possible to receive a continuous radio navigation signal at the same time that a two-way radio contact is being made with an aeronautical ground station. Such a unit is often referred to as a one and one-half communicator because it has a single transmitter and two receivers.

Most modern flying is done via the vhf omnidirectional range stations. These are called VOR or OMNI stations. In the vhf frequency spectrum, there is largely static-free reception, and the bending and false beams of the older low-frequency range stations are not present. A reliable directional pattern can be produced at these frequencies. A complex revolving antenna pattern that uses electronic switching generates a rotating beam that has a directional accuracy of 1° throughout the entire 360° of rotation. The aircraft need use only a simple nondirectional antenna.

More and more aviation radio gear employ an increasing number of integrated circuits. Reliability, compactness, and lightweight are the attractions.

The functional plan of a NARCO radiocommunication and navigation unit is shown in Fig. 10-20. Two antennas are connected to

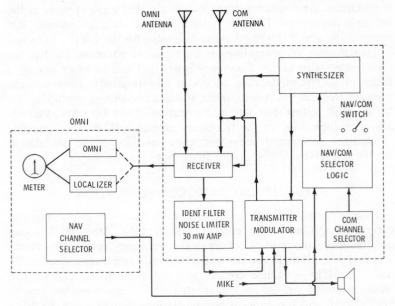

Fig. 10-20. Functional plan of aviation communications transceiver and omni receiver.

the receiver so it can be used for receiving communications signals, as well as the signals from the omnirange stations. When used for navigation, the output of the receiver is supplied to the omnisection on the left. Appropriate display meters are included, depending upon whether the signal is to be used for omnipositioning or for aircraft localization. It includes a channel selector which, through the main unit, sets up the proper oscillator frequency for the receiver.

In radiocommunication application, the receive signal is passed through an appropriate filter, i-f amplifier and detector, and audio amplifier prior to its application to the speaker. On the transmit position, a microphone input is activated. Voice signal is increased in level and then applied to the transmitter section of the unit.

A most important operation in the unit is the proper setting of receive and transmit frequencies. There are 360 individual channels involved, spaced 50 kHz apart. For ease of operation, this is done automatically by setting a digital readout dial that is calibrated in frequency. If the communication section is set to 119.5 MHz by the pilot, the unit must automatically set itself to this frequency, one out of a total of 360 possible frequencies. This work is handled by what is known as a *frequency synthesizer*.

The synthesizer responds to the NAV/COM mode selector, the NAV digital channel selector, and the COM channel selector. This logic information sets the transmitter on the desired frequency and the receiver for either omni or communications reception on the desired aeronautical radio channel.

In modern aviation radio equipment, it is the synthesizer that is designed around digital integrated circuits. Some ICs of various types are also found in other segments of modern aviation radio units. The array of digital ICs of the synthesizer can be seen in the photograph of Fig. 10-21.

Synthesizer Operation

It is to be anticipated that the frequency of 50 kHz is important in the operation of a synthesizer, because this is the channel spacing across the aviation radio spectrum. It is the reference frequency of

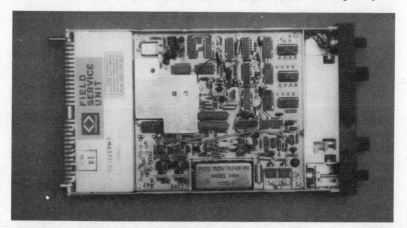

Fig. 10-21. Internal view of NARCO aviation radio showing array of ICs used in the synthesizer.

the phase-locked loop that is a part of the synthesizer, and it is at this frequency that a comparison is made in the phase detector (Fig. 10-22). The overall plan of the synthesizer is to generate output frequencies that are exact multiples of 50 kHz. However, these frequencies must be spaced in the high end of the vhf spectra. Hence, the indirect means of using a mixer and a vco must be employed in the generation of the spot frequencies that determine the transmitter output frequency and the receiver local-oscillator frequency.

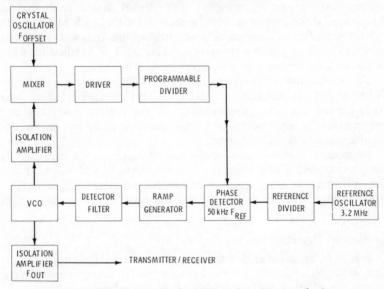

Fig. 10-22. Functional diagram of the frequency synthesizer.

In the NARCO radio, a reference oscillator operates on 3.2 MHz. It is divided by a factor of 64 to obtain the 50-kHz reference input for the phase detector. In the phase detector, the output of the programmable divider is compared with the reference voltage. The output of the phase detector is fed through a ramp-generator shaping circuit and the detector filter to provide a dc control voltage, which is applied to the voltage-controlled oscillator (vco). It is this oscillator that generates the output frequency that is passed to the transmitter/receiver isolation amplifier.

In transmit position, the output frequency selector sets the transmitter on the desired frequency. When the radio is switched to receive, there is a corresponding change in the vco output frequency. This new output frequency for local oscillator use is separated from the incoming receive frequency by an amount corresponding to the i-f frequency of the receiver. Thus, in switching between transmit and receive, the output frequency change corresponds to the i-f frequency.

The actual selection of frequencies by the aviation radio operator is handled by the programmable divider when he sets his digital dial to a desired frequency. Note that the phase-locked loop is completed from the vco to the mixer through an isolation amplifier. In the mixer, this component is compared with a crystal oscillator component known as the *offset frequency*.

The mixer function is to reduce the input frequency to the divider to an appropriate spot in the radio-frequency spectrum between 10 and 30 MHz. This frequency is now applied to the programmable divider. The programmable divider, in turn, divides this frequency by an amount determined by the setting of the selector switch, which controls the count of the programmable divider. In fact, this variable divider divides the input frequency by an amount that falls somewhere between 232 and 601, depending upon the selector setting. The selected divide ratio breaks down the input frequency to 50 kHz. It is this component that is applied to the phase detector for comparison with the reference frequency. If this frequency is more or less than 50 kHz, the vco frequency will shift correspondingly to make certain that the vco frequency is such that the division produces a 50-kHz output. Here we have the completed PLL loop.

The required division at the low-frequency end of the band would have to be higher than at the high-frequency end of the band. Let us use the count of 601. On the lowest channel, the desired vco frequency would be 118.00 MHz. The crystal offset frequency in the NARCO unit is 148.05. Therefore, the output of the mixer is 30.05 (148.05 − 118.00) MHz. The division by 601 produces a 50-kHz output only when the vco is guided precisely to 118.00 MHz by the phase-locked loop. This required count of 601 was established when the aircraft radio operator set his digital selector dial to 118.00 MHz.

The IC circuits of the NARCO synthesizer are shown in Fig. 10-23. At the bottom right, using discrete bipolar transistors, is the 3.2-MHz crystal-controlled reference oscillator that supplies signal to the 64-to-1 divider chain composed of U101, U102, and U103. From pin 6 of output gate G7, the 50-kHz reference pulse is supplied to the pulse phase detector, U104. The 50-kHz comparison output of the programmable divider is supplied to pin 13 of the same detector. Output is supplied to the succeeding ramp generator, which converts any phase difference to a corresponding dc voltage for controlling the voltage-controlled oscillator of the frequency synthesizer.

When changing channels and during intervals when the vco is not locked-in, an inhibitor component is supplied to the output system of the vco to prevent any vco component from being applied to the transmitter or receiver.

The difference frequency at the output of the mixer that heterodynes vco output and offset-oscillator output is applied to a driver.

The driver output is applied to the programmable divider, with IC U111 serving as a 2-to-1 divider and U112 as a control flip-flop. Output is then applied to U106, a 10-to-1 divider. However, through registry control it selects one appropriate pulse out of ten pulses applied to U112. This programming sets the 0.1- to 0.9-MHz channel. The total division is now 20.

Digital ICs U107 and U108, as a function of a selector, are required to divide by an appropriate integral between 12 and 29. This sets up the whole-number megahertz channels.

The 50-kHz separation is handled by the counting of the input control flip-flop, U112. Thus, the programming permits changing of the vco output in 50-kHz steps between 118 and 135.95 MHz as the selector dial is rotated.

PROJECT 7: WAVEFORM GENERATORS

General

In Project 7 you can construct a pair of waveform generators, checking out the capabilities of each. In Project 8 one generator is modulated by the other, forming either an amplitude-modulated or frequency-modulated output. The waveform generators use the Exar XR-205 IC (discussed in detail in Chapter 6) in conjunction with Figs. 6-6 through 6-10. Exar makes available in kit form two of these integrated circuits and a printed-circuit board for the generator. The complete schematic diagram is given in Fig. 10-24.

The generator is a source for the gamut of waveforms discussed in Chapter 6. These include sine, triangle, square, ramp, sawtooth, and pulse. The voltage-controlled oscillator can be either amplitude- or frequency-modulated. The waveform generator at the left of Fig. 10-24 can be used to modulate the generator at the right.

The one on the left has a fixed frequency, while that on the right has an adjustable output frequency by using range selector switch S1 and fine-frequency control potentiometer R31, center right. Beneath it is the output-amplitude control, R33. Potentiometer R22 at the lower left of IC2 is a waveform control. The two-section waveform selector switch (S2) is shown at the bottom right.

Potentiometer R10 is a wave-shape control for the fixed-frequency waveform generator IC1. Potentiometer R1 at the left center regulates the amplitude of the fixed generator output and is used as a modulation-level control. Potentiometer R6 at the top center also sets IC2 output level, but acts mainly as a balance control for proper adjustment of carrier level for conventional amplitude and suppressed-carrier modulations.

Waveform selector switch S3 is located at the bottom left. To its right is the modulation selector switch. Its center position is for off.

Position 1 uses the fixed-frequency generator on the left to frequency-modulate the variable-frequency generator on the right. Position 3 is used in the same manner for amplitude modulation. In Project 7 switch S4 should be set to its off position.

Construction

The two waveform-generating integrated circuits and the printed-circuit board can be purchased from Exar Integrated Systems, 733 North Pastoria Avenue, Sunnyvale, California 94086. Instructions and a parts list are included. After the circuit-board wiring was completed, it was mounted in a 5″ × 7″ × 3″ case (Fig. 10-25).

An oscilloscope is needed to check out the performance of the waveform generator. Additional components such as audio generator, phono player, microphone, frequency crystals, and a few capacitors are helpful, depending upon your specific needs for the waveform generator.

Construct the waveform generator. In this project you will check out the performance of each generator individually.

Operation

The schematic diagram (Fig. 10-24) shows the various switches, as well as the function of each switch position. Since the two generators are to be checked out separately, the modulation selector switch, S4, should be set to position 2 throughout Project 7.

The fixed-frequency generator on the left operates at approximately 1 kHz. Set the modulator waveform selector switch, S3, to its No. 1 (sine) position. Connect an oscilloscope across the modulation-output terminals (across resistor R36).

Turn on the waveform generator. Adjust and observe the output. The modulation-level control, potentiometer R1, is used to adjust the output amplitude. Observe the quality of the output waveform. It should be a good sine wave. The sine-wave-shaping potentiometer is R10. Adjust it for best quality output.

If you have an accurate audio oscillator, check out the frequency of the waveform generator, using a Lissajous pattern on your oscilloscope. This particular unit checked out at 1200 hertz instead of 1000. If you desire operation nearer to 1000 hertz, experiment with various other capacitors in the C2 position.

Set the waveform selector switch, S3, to the No. 2 (triangle) position. It may be necessary to make a slight readjustment of potentiometer R10 to obtain the very best triangular wave. Set switch S3 to the No. 3 position to observe a good quality square-wave output. Set switch S3 to the No. 4 (off) position.

Check out the variable-frequency generator, keeping switches S3 and S4 on their off positions. Transfer the oscilloscope to the signal-

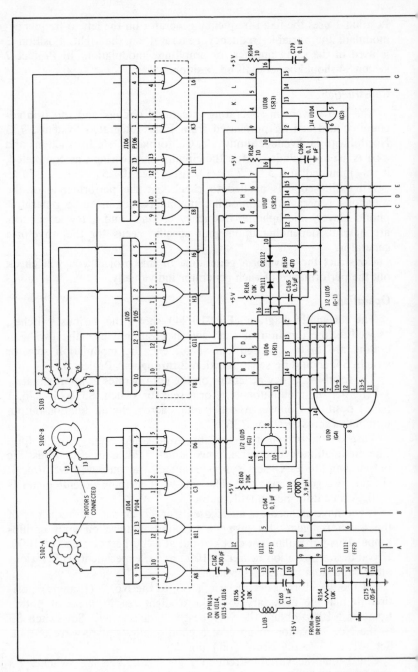

Fig. 10-23. The ICs in

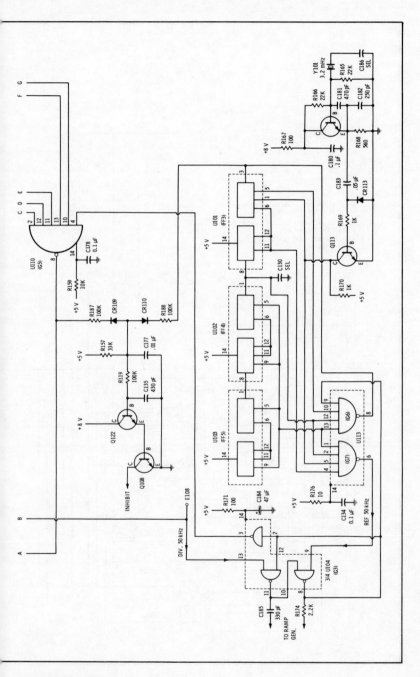

the NARCO synthesizer.

369

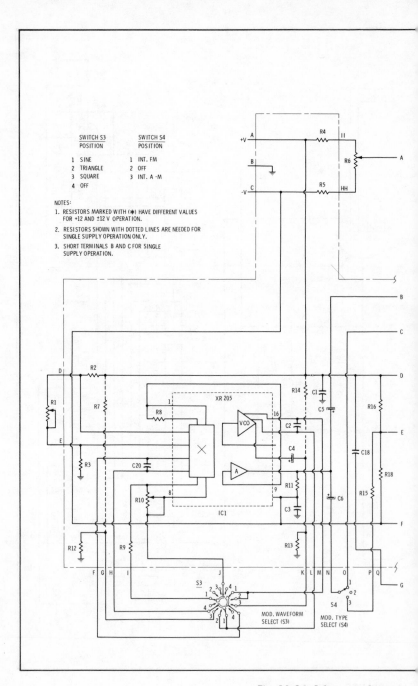

Fig. 10-24. Schematic diagram

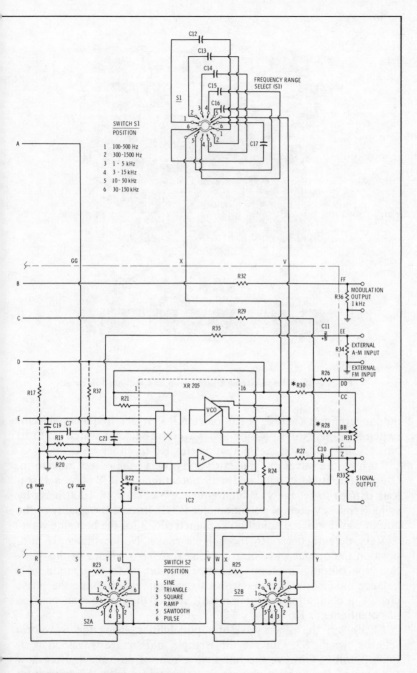

of modulated waveform generator.

Fig. 10-25. Printed circuit board wired and mounted in case.

output terminals (across potentiometer R33). Set frequency-selector switch S1 to its No. 3 position and waveform-selector switch S2 to its No. 1 (sine) position. Set potentiometer R6 to either end.

Turn on the waveform generator and observe the output waveform. Output amplitude is regulated with potentiometer R33. Use a Lissajous pattern to set the variable-frequency oscillator on 1000 hertz by using frequency-control potentiometer R31. Observe the quality of the output waveform and adjust potentiometer R22 for the best sine wave.

Vary potentiometer R6 through its range. Notice that near midposition the output waveform is reduced to zero. Hence at this position, the output waveform generator can be made to operate as a balance mixer or modulator with carrier suppression. Note the phase change in comparing outputs on each side of the zero-output position of potentiometer R6.

Check out the other waveform possibilities, using the waveform selector switch, S2. Observe the triangle and square waves on positions 2 and 3. Observe the ramp waveform of position 4. Note how the zero of the linearly rising voltage coincides with the zero ramp level. Move potentiometer R6 through its range, and notice how the direction of the linear voltage rise can be changed from one polarity to the other.

Observe the sawtooth waveform of position 5 of switch S2. Either a positive or negative sawtooth can be made available, depending upon the setting of potentiometer R6. Do the same for the pulse waveform at switch position 6. Negative or positive pulses can be made available, depending upon the setting of potentiometer R6. Reset switch S2 to No. 1 (sine) position.

Use an external audio generator to check out the frequency range of each position of switch S1. Approximate ranges for each position are given in Fig. 10-24. The actual range, of course, depends upon the absolute values of capacitors C12 through C17. These values can be adjusted to establish any desired frequency range. For example, a capacitor value of 2000 pF permits operation between 100 and 500 kHz. A capacitor value of 680 pF in any one of the switch positions would permit operation between 300 and 1500 kHz. The author inserted a 300-pF capacitor in one position to obtain operation over the entire broadcast band, providing a signal source for this entire spectra.

An advantage of this generator is its adaptability to meet a specific need. For example, a crystal socket was substituted for one of the S1 switch positions. Because crystals inserted into this socket produce a useful output above 10 MHz, you can set up spot frequencies to correspond to WWV signals. A small trimmer capacitor connected across the crystal socket will permit a precise WWV calibration. You may need a 1-MHz or 5-MHz crystal-controlled signal source. Perhaps your interest is spot frequencies on the 40-, 80- and 160-meter amateur bands. Remember that these frequencies can be either amplitude- or frequency-modulated.

Two changes can be made to provide added versatility. A crystal socket can be mounted externally and connected to one of the switch positions; thus, by plugging in an appropriate crystal, you can obtain any desired spot frequency. A small trimmer across the socket permits a fine adjustment of crystal frequency. Also, a pair of binding posts can be mounted externally and connected to one of the switch positions. An external fixed or variable capacitor can be inserted to establish preferred frequency ranges whenever they are needed. Capacitors of appropriate value can be connected across the binding posts to provide an output range of frequencies from very low to the upper high-frequency limit of the waveform generator.

The waveform generator also operates in the 9000-MHz range, the frequency used for generating sideband signals. Operation is also possible in the 8-MHz range, customarily used for generating frequency-modulated signals.

PROJECT 8: MODULATED WAVEFORM GENERATOR

General

In this project a modulating link is established between the wave-form generators. The fixed-frequency oscillator is used to amplitude-modulate or frequency-modulate the variable-frequency oscillator. Position 1 of switch S4 provides frequency modulation; position 3 provides amplitude modulation. The actual modulation level is controlled with potentiometer R1, which sets output amplitude of the fixed-frequency generator. Potentiometer R6, of course, is used as a balance control, providing carrier cancellation for DSB and carrier level control for conventional a-m.

A high-impedance, high-output microphone or a phonoplayer can be used as an external source of modulating signal. The oscilloscope is again used to display the waveform. A broadcast receiver can be of help in listening to the modulated output of the waveform generator. In fact, the combination can serve as a wireless record player.

A microphone and waveform generator can be used as a versatile amateur test signal or as a source of signal for a transmitted signal, provided amplification is included. Linear amplification is needed to build up the weak double-sideband or conventional a-m modulated output of the generator. However, conventional class-C frequency multipliers and amplifiers can be used to build up any frequency-modulated signal made available by the waveform generator.

Operation

Leave selector switches S3 and S4 on their off positions. Set frequency switch S1 to the No. 6 position. Set the waveform selector switch to position 1 (sine).

Turn on the waveform generator and adjust the frequency-control potentiometer R31 to generate a 100-kHz output. Use a Lissajous pattern on your oscilloscope. Adjust for a good quality sine wave, using potentiometer R22. Adjust potentiometer R6 for complete carrier suppression.

Turn on the fixed-frequency oscillator by setting switch S3 to its sine position. Apply a-m modulation to the output waveform generator by setting selector S4 to position 3.

Position a broadcast receiver near to the waveform generator. Note that the signal can be heard every 100 kHz over the broadcast band. If your radio has a short-wave band, notice that it can also be heard at closely spaced intervals over the short-wave spectrum.

Momentarily turn off the modulation by setting switch S4 to position 2. Use potentiometer R6 to generate some carrier in the output. Now set switch S4 to the a-m position again. Adjust potentiometer R6 and modulation-level control R1 to obtain a good quality 100 percent amplitude-modulated waveform at the output. Note that a very excellent a-m modulation pattern can be obtained. Notice that this too can be heard over the broadcast and short-wave bands.

Set switch S4 to the off position. On the oscilloscope screen, display four cycles of the 100-kHz wave. Now set the modulation selector switch to position 1 (fm). Note there is a slight fuzziness at the peak of each waveform when the modulation-level control is set to maximum. Display only one cycle of this wave, and vary the modulation-level control between minimum and maximum. Now it is possible to see the displacement of the waveform with 1000-Hz fm modulation.

This signal, too, can be heard on the broadcast band, along with additional ringing, which indicates beats between the various sidebands of the generated fm signal. A communications receiver with very narrow band-reception capability, such as used for receiving cw signals only, will be able to distinguish between the carrier and the multiple sidebands generated by an fm signal.

If you have included the 300-pF broadcast-band capacitor, set your frequency-selector switch to this position. Locate an empty spot on the a-m broadcast band. Adjust the frequency-control potentiometer, R31, until you tune in the signal. Adjust modulation-level control R1 and carrier control R6 to obtain an ideal 100 percent modulated a-m signal. At these higher frequencies, use switch S2 on position 3 to obtain the best modulation pattern.

Turn off the internal modulation by setting switch S3 to off position. Connect a microphone to the a-m input of the waveform generator. Speak into the microphone, and listen to the quality of the signal picked up by the broadcast receiver. It may be necessary to set potentiometer R6 to a lower carrier level to obtain a high level of modulation. If one is available, substitute a phonoplayer and listen to the music quality as reproduced by the broadcast receiver.

If you have inserted a crystal position, check out the performance on the various crystal frequencies by using the internal modulation. If you are a licensed radio amateur, you may wish to use a 160-meter crystal to generate an a-m signal or a double-sideband and suppressed carrier on 160 meters. Also, it would be possible to frequency-modulate a somewhat higher-frequency crystal and obtain an output on the 2-, 6-, or 10-meter fm bands by using a chain of frequency multipliers.

Index

B

Balanced
  mixer, 143
  modulator, 154, 172
  transistors, 137
Bandpass amplifier, 120
Barrier potential, 16-17
Base
  current, 25
  junction, 20
Basic
  audio stage, 235-237
  class-B output stage, 87
  differential amplifier, 67-69
  fabrication methods, 41-42
  IC electronic voltage regulator, 181
  IC structures, 7-10
  logic blocks, 217-220
  monolithic IC, 40
  motor-control methods, 318
  operation of op amp, 109-112
  operational amplifier applications,
    121
  transistor flip-flop circuit, 225
Bias
  constant-gain, 167
  constant voltage, 167
Biasing, 21-22
  modes, 167
Binary
  -coded-decimal (BCD), 214
  numbering, 213-215
Bipolar transistor, 19-25
Block diagram of voltage-regulation
    system, 180
Broadband video amplifier, 174

C

Calculating section, 216
Capacitor, monolithic, 54
Capture range of PLL, 202
Carrier motion, 14
Carriers, majority and minority, 25-28
Cascaded ICs, 144-145
Charge motion, 15-16
Chroma processing and demodulation,
    291-299

Circuit(s)
  COS/MOS-oscillator, 353
  demodulator, 351
  ECL, 222-223
  level-shifting, 83-85
  logic, 215-217
  nulling, 117
  output, 85-88
  thermocouple response, 314
  TTL, 220-229
  vox, 344-345
Clocked S-R flip-flop, 227
Collector cutoff current, 27
Commercial applications of ICs,
    358-361
Common-mode
  input voltage, 76
  rejection ratio, 76
  voltage gain, 76
Comparison of discrete and IC
    transistor, 45
Components, passive, 51-56
Connecting an IC for high gain, 147
Constant
  -current sources, 77-80
  -gain bias, 167
  -voltage bias, 167
Control system, motor-speed, 322
Controller
  motor-speed, 321-323
  temperature, 312
COS/MOS
  array, 350-353
  crystal-oscillator circuit, 353
  linear amplifier, 354
Counter, frequency, 332
Current
  avalanche, 19
  differential amplifier, 72-73
  diffusion, 24
  in a semiconductor, 14
  transistor, 23-25
Curves, FET characteristic, 31-33

D

Darlington configuration, 70-71
Data input, 216
Dc voltmeter, high-impedance,
    325-327

378